JOHN MURR ˉTICE

teaching seconaury
SCIENTIFIC ENQUIRY

University of
Chester

teaching secondary
SCIENTIFIC ENQUIRY

Editors: David Sang
Valerie Wood-Robinson

JOHN MURRAY

Titles in this series:

First published in 2002
by John Murray (Publishers) Ltd, a division of Hodder Headline Ltd,
338 Euston Road
London NW1 3BH

Illustrations by Art Construction
Layouts by Amanda Easter

Typeset in 12/13pt Galliard by Wearset Ltd, Boldon, Tyne & Wear
Printed and bound in Great Britain by Alden Press, Oxford

A catalogue entry for this title is available from the British Library

ISBN 0 7195 8618 6

Contents

Contributors

Peter Campbell has taught Physics in London comprehensive schools for over 20 years. During 2001–02 he managed a joint project for the ASE and The Wellcome Trust, developing materials and preparing teachers for teaching Citizenship through Science at Key Stage 3. He has recently joined the Nuffield Curriculum Projects Centre where he is working on the 21st Century Science project.

Linda Ellis was a Science teacher and Head of Science in Leicestershire before moving to advisory work. She was Science Adviser in Hertfordshire and is now working in Brighton and Hove. She was part of the writing team for Warwick Process Science and has contributed to ASE INSET courses around the country.

Peter Ellis has been teaching Science for over 25 years and is currently engaged in a number of educational publishing ventures. He is interested in the use of the history of science in teaching and is a member of the British Society for the History of Science Education Section committee.

Mike Evans was a Science teacher and Head of Science in a number of state schools before moving into advisory work. He was Science Adviser in Hampshire and then in Southampton and since 2001 has been employed by the Centre for British Teachers. He has written science text books and contributed to ASE publications and courses.

Richard Gott works at the University of Durham. His research and writing interests revolve around practical work in science and issues related to the public understanding of scientific evidence.

Nigel Heslop has been a Science teacher for 29 years, 20 as Head of Department. He is now an international consultant for Science curricula and learning through language, particularly for Eastern Europe. He has written for many publishers, most recently the Hodder Science KS3 series. He was part of the ASE team producing CD-ROMs and providing training for Science Year and for Planet Science.

Mary Ratcliffe taught Science in comprehensive schools in East Anglia for over 15 years before joining the University of Southampton where she is currently a senior lecturer in Science Education. She has research interests in teaching and learning of socio-scientific issues and has written extensively in this field for different audiences.

Ros Roberts taught in comprehensive schools before joining the Science Education team at Durham University. Her teaching and research interests include Biology, Biology Education and teaching about scientific evidence.

David Sang worked as a research physicist for 6 years before becoming a teacher. He has taught Physics for 13 years, in a comprehensive school, a sixth-form college and a polytechnic. He now works on a variety of curriculum development projects, as well as writing text books and multimedia teaching materials for pupils aged 11 to 18.

Tony Sherborne works at the Centre for Science Education, Sheffield Hallam University. He is project leader for the Pupil Researcher Initiative, and develops paper and multimedia resources to teach contemporary science. Previous work includes Head of Development for ASE Science Year CD-ROMs, teaching for 7 years and working for BBC Education.

Rod Watson taught in schools for 13 years before joining King's College London, where he is now a senior lecturer in the Department of Education and Professional Studies. He has carried out research and development in the area of scientific enquiry over 15 years and was the Director of the ASE/King's Science Investigations in Schools (AKSIS) project. His publications relate to teaching in schools, professional development and research.

Valerie Wood-Robinson has been Head of Biology, Head of Science, and a Science Adviser. She has interspersed her teaching with work on research and curriculum development projects, most recently the AKSIS Project. Allegedly retired, she is still busy writing and tutoring ITT students.

Acknowledgements

Author's acknowledgements

This book represents the efforts of many people. The editors are grateful to all of the contributors, and in particular to Mike Evans, who first suggested that we devise this book. We are also grateful to Martin Hollins (Qualifications and Curriculum Authority) and Daniel Sandford Smith and Jane Hanrott (Association for Science Education) for their support and guidance during the development of the book.

The Association for Science Education acknowledges the generous financial support of ESSO and the Institution of Electrical Engineers (IEE) for the ASE/John Murray Science Practice series.

Photo credits

Photographs are supplied courtesy of:

Cover Mehau Kulyk/SPL; **p.22** Robert Brooke/Photofusion; **p.69** Trevo Perry/Photofusion; **p.143** Mary Evans Picture Library; **p.158** Imperial Cancer Research Fund.

The publishers have made every effort to contact copyright holders. If any have been overlooked they make the necessary arrangements at the earliest opportunity.

Introduction

David Sang

This book joins the ASE/John Murray Science Practice series as an accompaniment to *Teaching Secondary Physics*, *Chemistry* and *Biology*. Whereas those three previous volumes looked at ways of teaching the subject content of science courses in secondary schools, this book looks at a complementary aspect of science: scientific enquiry.

Scientific enquiry considers two important, related questions:

- How can our students carry out their own investigations in a scientific way?
- How does the scientific community develop new ideas, supported by empirical evidence?

We live at a time when the authoritative pronouncements of scientists are no longer automatically accepted by citizens. Many young people turn away from science, because they see it as responsible for many problems in the world (nuclear weapons, pollution and BSE, to choose just one each from the three major branches of science which they study), and they consider that science has little to say about questions that they regard as important. At the same time, scientific advances and the technologies that flow from them are having an ever-growing influence on our lives.

Hence it is important that young people should have a clearer grasp of the nature of science and how scientists work. They need to be able to look at scientific evidence and assess for themselves how strongly it supports the ideas that are being developed from it. They can learn how to do this by examining current and past controversies involving science, and by developing their own skills of enquiry.

The Science National Curriculum in England and Wales now reflects a clearer view of the way in which our students can develop their ideas about scientific enquiry. At the same time, the Scottish curriculum has moved in a similar direction. Several of the authors represented in this book have been closely involved in these developments. Although the National Curriculum spells out the strands within scientific enquiry and the sequence in which students are likely to develop their skills, this book is intended to help teachers devise schemes of work which will help students to develop these skills.

The first part of the book looks at 'scientific investigation'. Here, we are most concerned to develop students' understanding of the nature of scientific evidence (their 'ideas about evidence'), and how this can guide them through investigative work. The second part of the book looks at 'ideas and evidence', with a greater emphasis on how students can approach scientific ideas from beyond the school laboratory.

This has been a harder book to write than the three earlier volumes in the series. It is relatively easy to say, 'here is a good way to teach about electromagnetism, or human genetics'. To practising teachers, such topics are familiar, and they find it easier to incorporate new ideas into their existing schemes of work. The purposes of practical and investigative work have, of course, been discussed over the decades. However, the present emphasis on the nature of scientific evidence is relatively new and unfamiliar to most of us. (Most science teachers have not been practising scientists.) Hence we have tried to devise a helpful route map through this territory, referring to readily available resources wherever possible.

At the end of this, we believe that your students will have a better grasp of the nature of scientific enquiry. As future scientists and as future citizens, this can only be of benefit to them and to the wider community.

1 Investigations: Introduction

Valerie Wood-Robinson and members of the author team

Scientific investigations are about collecting evidence. For students to carry out a successful investigation they need to have some clear ideas about the nature of evidence. How good is the evidence they have collected? How can they judge its value? What explanation does their evidence support? What explanations can be ruled out?

1.1 The need for teaching

Which of these two quotes is typical of your students?

> 'Please sir, I've been trying really hard for twenty-five minutes and I've only written the title. I've got no idea where to start.'

> 'Now we know how to write a better conclusion, looking at evidence from our results. We couldn't have done it before because we didn't know what they (the teachers) wanted.'

The second student is clear about what is expected in an investigation, because he/she has been taught this explicitly. The first has not had such explicit teaching.

To put it more formally: 'Understanding of empirical evidence and of criteria for evaluating the quality of evidence needs to be explicitly addressed through the curriculum.' (Millar *et al.*, 1994).

This is the keynote to the next four chapters. The focus is on evidence when teaching about investigations. Note that here we are talking about *ideas of evidence* – in other words, ideas that we want our students to use when thinking about scientific evidence in the context of their own investigative work. (In Chapters 6–10, we will consider the part of the science curriculum known as *ideas and evidence*, which looks at how ideas and evidence are interrelated within science in general.)

The ideas of evidence we are talking about here, as with all ideas, can be taught, and put to use, in a variety of contexts. They can be taught and assessed using both practical and non-practical work. The important point is that they need to be *explicitly taught* as well as being assessed. Some students will pick up these ideas in the course of studying the traditional content of science, but many will not.

Of all the attainment targets in the National Curriculum for Science, Sc1 (now called 'scientific enquiry') has been the most neglected in the early years of secondary school. Skills are taught to pupils in primary schools, but because teachers perceive the secondary curriculum as crowded with factual content, little time is spent on the development of investigative ideas. Assessment of investigative work for GCSE dominates Key Stage 4 and often pervades Key Stage 3, without the teaching to underpin it.

Here is what Ofsted said about this in the report *Progress in Key Stage 3 Science* published in March 2000:

> Experimental and investigative science (AT1) is mainly used as an assessment activity and has not yet become an integral part of science teaching at Key Stage 3. When well taught, this section of the Programme of Study motivates pupils and raises standards. . . . Experimental and investigative science has led to some very high quality practical work but only by a minority of pupils. It has not become an integral part of secondary science teaching. All too often at Key Stage 3 it is relegated to an assessment activity, bolted on to the rest of the curriculum, mainly in Year 9. Where it has been incorporated in a more integrated way into the science curriculum it has given a sense of purpose to laboratory work and proved motivating to all pupils and especially the most able. . . . When it is included in the science curriculum it is often as an assessment activity judged using GCSE examination criteria; these encourage a mechanistic approach. Planning and considering the strength of evidence are often neglected in routine practical work.

We hope that the discussion and teaching suggestions in the following chapters will help teachers address these neglected areas and assist integrated understanding of the ideas of evidence.

In Chapter 2, Richard Gott and Ros Roberts introduce the 'ideas of evidence' which make up the 'toolkit' for scientific enquiry. They outline a real-life problem as the context for teaching suggestions to teach the ideas involved in measurement, design and reporting, that is 'collecting and using evidence'. The terminology of 'ideas of evidence' has been continued through the next three chapters, though we have sometimes used the term 'skills'. However, we want to avoid this suggesting a view of routine or mechanical skills. The skills of enquiry involve a lot of thinking, not mindless routines.

In Chapter 3, Nigel Heslop and Valerie Wood-Robinson offer further suggestions for teaching the planning of investigations, linking with evaluation and developing the idea of 'planning from questions'.

In Chapter 4, 'Considering evidence' Mike Evans and Linda Ellis give suggestions for teaching the ideas of relationships, conclusions and the link to scientific concepts.

Chapter 5 consolidates the ideas of 'evaluating evidence'. Valerie Wood-Robinson and Rod Watson suggest ways of judging the quality of evidence, with teaching activities that link back to questions of measurement and design from Chapters 2 and 3.

The chapters are arranged in a traditional sequence with evaluation coming last, but this is not to suggest that the teaching of scientific enquiry can or should follow this linear sequence. As Richard Gott and Sandra Duggan have pointed out:

> The evaluation of the task requires an understanding of all . . . stages (of the investigation) . . . , and thus understanding of evaluation is needed as much at the beginning as at the end of the task.

The iterative nature of the investigative process, and of teaching it, is emphasised by profuse cross-referencing between chapters.

The ideas of evidence are universal to understanding scientific enquiry. The content is directly relevant to teaching the National Curriculum in the years of statutory secondary education in England and Wales, but is not directly referenced to the statutory Order. The suggestions are not divided into age groups; in places we have referred to younger or older students within the secondary age range. They are suggested activities, not a course. Teachers will want to select appropriate ideas and activities to allow for differentiation and progression. None of the suggestions will produce a 'quick fix' to instant learning. Some ideas of evidence are difficult to assimilate and will need revisiting and reinforcing. Most of the teaching suggestions are strategies for focused teaching of specific ideas. Suggestions for assessment are similarly focused on specific ideas, not on 'whole investigation' summative assessment.

♦ *Progression and assessment*

When we think about progression in investigations, what is it that is to progress? Richard Gott and Ros Roberts in Chapter 2, page 19, suggest it could be the level of *manual dexterity*, the sophistication of the *context* (e.g. from sugar with measuring cylinders to potassium permanganate with a burette) or the sophistication of the *explanatory ideas* (particle theory, photosynthesis, etc.) that underlie the experiment. In these chapters we shall refer to progression in the *underlying ideas* related to design, measurement and analysis.

In addition we can think of progressing literacy skills related to enquiry. Students' thinking also progresses from an intuitive sense of ideas like fairness and suitability to a more formal understanding of the ideas of evidence with their correct words.

Your students' performance in the suggested activities will contribute to diagnostic assessment; you will find out how well they understand the ideas and probably be made aware of their misunderstandings. Familiarity with students' difficulties and misconceptions helps teachers to focus their teaching on alleviating problems:

> As a medical practitioner diagnoses the cause of a symptom before attempting to alleviate it, so the teacher needs to diagnose the viewpoints of his/her pupils before deciding how to set about modifying them towards more scientifically acceptable ones (Osborne & Freyberg, 1985).

The knowledge and experience of most students prior to starting secondary school, and some of their common problems of understanding, are listed at the start of each chapter and in some of the activities. Ofsted report that standards of achievement at the start of secondary school are higher than previously as a result of improvements in primary science. Students show improved ability to use simple scientific terminology correctly, take careful measurements of length, volume, time and temperature, record and display measurements and observations, make decisions about how to turn simple questions into experiments and recognise that a 'fair test' means keeping conditions the same. That gives a springboard for refining students' ideas of evidence.

The stages of individuals' progress may be similar to the range of understanding among students within one class. Thus progression of differentiation are two dimensions of considering different levels of understanding that children may have. Monitoring the progress of individuals is a component of differentiation.

Many of the strategies and examples are equally applicable throughout the secondary school, especially as they can be used as individualised (differentiated) approaches. Indeed some of the activities can be reused as diagnostic exercises to see how thoroughly students understand the ideas and where further development is needed.

♦ *Terminology*

There is a lack of consensus over the definition and usage of some words used in enquiry, both among professional scientists and among science teachers. We have tried to use consistent terminology here and to discuss issues relating to the best words to teach students to use. It is advisable that all teachers within a school use the same terms, share the same meanings, use the terms consistently and avoid aggravating students' misconceptions. This could be written into departmental policy and schemes of work, together with proposals for teaching ideas of evidence.

1.2 Teaching strategies

The same principles and range of strategies apply to the teaching suggestions in all of the subsequent chapters.

Teachers should not expect students to do a whole investigation *thoroughly* before the latter have the knowledge, overview and techniques to cope with it. Students need to be taught directly the component ideas and skills by explicit teaching focused on one idea at a time. Teaching investigations is an iterative aspect of the business of enquiry. With a new class you might like to let the students 'have a go' at an investigation from their prior experience. This will give you an idea of their areas of strength and weakness. Then with an activity focused on, say, evaluation of the sufficiency of data, students will understand about the quality of the evidence they need to collect. They can then apply what they have learned when engaging in another scientific enquiry. You will need to revisit component ideas as you embed them in 'the whole picture', to clarify, reinforce and progress them. This may appear very time-consuming, but you can do this by providing many and regular short opportunities, particularly in advance of sessions where you want students to apply a particular idea in performing a scientific enquiry. These short sessions could be whole class starter sessions, homeworks that are then discussed in class, or group work that results in both students and teachers discussing the principles involved.

In order to teach these ideas and skills you will need access to a range of sets of data. You can make these up yourself, use those generated by students from previous years, or get them from scientific journals or the internet. Real data or 'messy data' (such as air quality data from the internet) can provide real opportunities to discuss anomalous data. Resources for teaching these ideas are suggested at the end of this chapter and throughout Chapters 2–5.

Successful teaching of ideas of evidence depends on variety of teaching styles, in both practical and non-practical lessons. These include direct teaching of examples, interaction with students individually and in groups as they perform their enquiries, and various strategies for active learning. The range of strategies listed here and suggested in the following chapters is not exhaustive, however, and both well-tried teaching techniques and innovative new ones can be used.

◆ *General points*

When devising teaching activities to develop your students' investigative ideas, bear these points in mind:

- The learning objective of the lesson (e.g. to evaluate the sufficiency and accuracy of the measurements) is different from the aim of the enquiry (e.g. to find how the mass of a vehicle affects the distance it travels). Students need to distinguish between these, so you should clearly explain the learning objectives in terms of learning how to do certain aspects of enquiries and investigations.
- The activity should involve active engagement of the students, whether in practical experimental work that exposes them to a new technique or piece of apparatus, or in a non-practical activity that exposes them to new ideas.
- The activity should use the target vocabulary about science investigations and emphasise relevant text types (link with literacy).
- If possible, there should be a challenge that makes the students think critically about the viewpoint or model they are using for that task. This approach is used in the *Thinking Science* (CASE) materials.

♦ *Examples of strategies*

This list is not exhaustive. It does not include all the strategies referred to in the following chapters, for example brainstorming, question and answer discussions or poster presentations, which are familiar to teachers as general strategies. Many of these strategies are suitable to use in whole-class starter sessions or as homework exercises that are later discussed in class or group work. In addition, such strategies as drama, role play, simulation, active reading, and writing for different audiences are also explained in detail in the case studies of ideas and evidence in Chapters 7, 8 and 9.

A. Using exemplars to imitate or criticise

♦ **Modelling** of examples to emulate is a very good way of showing students what is expected and helping them understand what is meant by quality of planning, generalisations, evaluation, or whatever you are teaching. You could collect together some examples (e.g. from previous students' work and from scientific writing) or write your own. You could then read through some, as a class, pointing out the good features. This strategy is suggested in the following sections:

- a linguistic convention, 'As this, so that', comparative adjectives (see teaching suggestions 1, 3 and 5 on pages 76–80)
- good write-ups (see teaching suggestions 5 on page 46 and 1 on page 69)

♦ **Criticising others' statements** is a strategy similar to modelling exemplars, but in this case you include examples with shortcomings, usually common mistakes. Students enjoy 'playing teacher' if asked to 'mark' work. They accept the statements as realistic, particularly if the work is identified by (fictitious) students' names. They find it more comfortable to criticise the work of others than their own or their classmates' work. One way is to get students to rank statements in order of quality, identifying their good and bad points. A suggested strategy is that you ask students to work in small groups to discuss and order the statements, then to compare and justify their decisions with other groups in the class. Whenever you present examples with shortcomings, make it clear to students that they are *not* presented as models to emulate.

Although you may want to present exercises as pencil and paper or discussion tasks, you will want your students to understand the context from which the examples are derived. You may need to do a quick demonstration to make the context clear. This strategy is suggested in several sections (see teaching suggestions 2 on page 68, 6 on page 80, 3 on page 85, 3 on page 86, 2 on page 93, 1 on page 99, 1 on page 100 and 1 on page 102).

♦ **Marking** a class partner's actual work, rather than examples contrived for the purpose, is an opportunity for formative assessment. It is important to stress the need for *positive* criticism. Students should be asked to comment on focused sections of work (e.g. describe a plan, or explain anomalies) rather than mark a whole investigation write-up. This is illustrated in teaching suggestions 1 on page 67 and 3 on page 86).

♦ **Improving someone else's writing:** Having criticised the invented or real examples, a task for students is to try to write a better version themselves, as suggested in teaching suggestion 2 on page 71.

B. DARTs (Directed Activities Related to Text)

♦ **Text marking** includes underlining, colour coding, annotating or numbering items in text, to find specific ideas and categories of information. It is suggested in:

- yeast text (teaching suggestion 2 on page 53), photosynthesis text (teaching suggestion 2 on page 56), freezing water text (teaching suggestion 1 on page 69)
- identifying relationships being investigated (teaching suggestion 3 on page 45)
- identifying component sources and misrepresentation (teaching suggestion 6 on page 43)
- evaluating student results (teaching suggestion 4 on page 96).

♦ **Text restructuring** involves reading, then remodelling the information in another format. This is suggested in rewriting plans (teaching suggestion 6 on page 64) and could be applied to rewriting text from the press (teaching suggestions 5 on page 43 and 1 on page 106). Similarly, the information expressed in written descriptions, tables and graphs can be interchanged, as in teaching suggestions 2, 3 and 5 on pages 77–80.

♦ **Cloze technique** (i.e. filling in the gaps in text with your own words or from a list of given words) and **sequencing** (i.e. reconstructing text that has been cut into chunks) can be used for learning to write coherent plans, conclusions and evaluations.

♦ **Card sorting**, using questions, words or statements written on sets of cards, saves students the task of laborious writing. It focuses immediately on the content in the form of a game. Students match pairs of cards, or sort them into sets, working as pairs, groups or snowball groups. You can also incorporate cards into other strategies: for cloze, sequencing, role descriptions, scenarios, etc. Preparing colour-coded cards and keeping them in reusable packs saves time in preparing consumable worksheets.
Examples include:

   • sorting questions into kinds of enquiry (teaching suggestion 1 on page 53)
   • matching generalisation and conclusions or explanations (teaching suggestions 2 on page 83 and 4 on page 85).

C. Using simulations

♦ **Models** or small scale simulations are useful where real space and time preclude real practice (e.g. for sampling); examples are a 'battleships' game (teaching suggestion 3 on page 40), newspapers and cover-slip quadrats (teaching suggestion 4 on page 42). The processes of enquiry can also be modelled by computer programs that simulate whole or part investigations (see below for ICT-based strategies).

♦ **Role-play simulations** can provide a realistic context for oral presentation, discussing and justifying evidence. You can get students to present their own, or second-hand, results and conclusions to the rest of the class, who then criticise and quiz them. This approach could be used in teaching suggestions 1 on page 86, 2 on page 104, and 1 and 2 on page 105). The presenters have to justify their results against criticism which helps them to focus on the interpretation of evidence and possible weaknesses in their evidence. Role play allows students to argue their case and accept criticism in their role, as distinct from their real egos, for example in evaluating evidence for and against vaccination, burning waste, siting phone masts, etc.

A simulated scientists' conference, public enquiry or pressure group meeting could take place in a normal lesson or be elaborated for an assembly, open day, science fair or science week activity. This is suggested in teaching suggestion 4 page 55 and could be used in teaching suggestion 2 on page 85. It could be based on the 'Chimney' case study used throughout Chapter 2.

♦ **Adopting writing styles** in different roles (such as journalist, irate neighbour or scientist) and for different audiences, is familiar to students from their literacy studies. You can use this strategy to challenge students to use their ideas of evidence in justifying their evidence to classmates, younger children, newspaper readers or others. This helps students to disengage the ideas from a formulaic 'write-up' style and to engage them with creative thinking.

♦ **Concept cartoons** enable students to identify with the supporters of alternative theories. They provide instant and attractive presentations of competing theories, including the 'misconceptions' that students themselves are likely to hold. They provide a stimulus for considering the evidence needed to support competing theories (teaching suggestion 1 on page 85).

D. ICT-based activities

The Science Investigator (see Other resources, page 16) simulates moving stickers of variables, on posters. Starting from identifying variables and designing an investigation, it follows through to recording, analysing and evaluating (teaching suggestion 3 on page 63).

The *Getting to Grips with Graphs* disk (see Other resources) contains several interactive tasks, including 'Link the actions'. This gives practice in expressing relationships in a context-free environment (teaching suggestion 1 on page 76).

An internet search is useful for performing enquiries that cannot be carried out in person (teaching suggestion 1 on page 56). It can be used as a source of scientific knowledge to explain conclusions. However, students should be advised against copying information that they do not understand and that has little relevance to their own explanations.

'*Building Success in Sc1: Science Investigations for GCSE*' (see Other resources). This is a versatile multimedia resource that can be used with any scheme of work. It uses a real bank of data on a CD-ROM and can be used to teach the ideas of Sc1 as well as providing large data sets that cannot always be achieved in school practicals. It is supported by a copiable teacher's guide.

♦ **Virtual experiments** using Focus Educational *Science Investigations* enable students to work through the processes of investigations. This software includes a number of investigations where users can change the values of variables by manipulating diagrammatic apparatus, run the virtual procedure and take readings from scales. They have to go through the thinking and decision-making of hands-on investigations, enhanced by the possibility of numerous repeats and reruns in a short time. Free sample investigations are available from their website. (See Other resources, page 17.)

1.3 Prompt sheets and writing frames

♦ **Prompt sheets** (help sheets) are commonly used in teaching science investigations. A prompt sheet can be worded as a checklist, as a series of questions in 'student-speak' or as a Cloze exercise. You need to be clear about the purpose of your prompt sheet and design it for a specific use – whether as a prompt for action, scaffolding for thinking, or as a writing frame to model the structure of phrases to express ideas. One design of prompt sheet rarely fits more than one purpose. You may need to make a prompt sheet specific to a particular enquiry. If you use a generic one that suits a range of investigations check that it is appropriate to the investigation in hand. This is particularly so when prompts about, for example, controlling variables and fair tests may not be relevant to the kind of enquiry being undertaken.

Ideally a prompt sheet serves as scaffolding, to be removed from use when the structure is established, and not as a crutch that cannot be discarded or, even worse, a straitjacket that constrains development. The potential limitation imposed by a prompt sheet is indicated in advice from the GCSE examination boards. They suggest that the use of a prompt sheet which is very closely related to the context of the investigation severely restricts the scope for the student to make decisions in each of the Skill Areas and this inhibits access to higher marks in each of the Skill Areas. They recommend that prompt sheets of this type may be suitable during the early stages of teaching and learning of the requirements of Sc1, but teachers are advised to use generic (context-free) prompt sheets when investigations are assessed for GCSE purposes.

Some school departments design a generic detailed prompt sheet as complete scaffolding to start students in their investigations. Then they progressively take away particular parts from subsequent print-offs, as classes and individuals progress to need less support. Another variation is to ask students to prepare their own checklists, or to prepare checklists or prompts suitable for younger 'beginner investigators'. You could also ask for feedback from students about how helpful they found the support sheets; let them know that they are meant to be helpful tools and are not 'tablets of stone' inherent to scientific enquiry.

Scaffolding strategies for whole class or individual use are mentioned throughout Chapters 2 to 5, including several interactive strategies particularly in Chapter 3, section 3.4. When scaffolds serve the purpose of writing frames the link with the literacy concept of 'text type' is enhanced.

◆ *Prompting planning*

Box 1.1 is an example of a prompt sheet for planning.

Box 1.1 *Planning an investigation: prompt sheet*

In science lessons, the topic I have learned about is ....

The aspect that I am going to investigate is..........................

The question I want to ask is ....

The variables that affect this question are ....

My input variable will be ....

My output variable will be ....

I will keep these variables the same ....

What I am going to do is ....

The apparatus I will use is ....

I am going to make this a fair test because ....

What I expect to find out is ....

I think this will happen because ....

A more extensive prompt sheet for thinking about planning would include questions relating to range, the number of values and adapting from a trial run, and, in addition, planning for fairness, accuracy and repeatability where these are relevant to the particular investigation.

◆ *Prompting considering*

Boxes 1.2 and 1.3 are two examples of writing frames for 'Conclusions'.

Box 1.2 *Concluding frame 1*

Looking at my results I noticed that as .... then ....

This was expected/not expected because ....

The reason for the pattern is ....

So my conclusion is ....

Box 1.3 *Concluding frame 2*

I have been investigating ....

The results table headings are ....

Plotting a graph of (?) .... against (??) ....

will show the connection between (?) ....

and (??) ....

The shape of my graph is ...., which shows me

that as .... so ....

My conclusion is ....

The explanation for this is ....

♦ *Prompting evaluation*

The process of investigating is continuous and iterative, so a prompt sheet should not confine thinking and explaining to distinct sections. The scaffold in Box 1.4 starts by directing the student to consider the pattern of results, to return to planning, and then to evaluate.

Box 1.4 *Evaluation: prompt sheet*

From my series of results, I can see this pattern ....,

which suggests ....

I think this happened because ....

The following results do not fit the pattern ....

Having repeated these and/or other values, I got results that fitted the pattern/I got the same results that did not fit the pattern.

I think .... because ....

I think the data I collected may/may not have been

sufficient to prove my point because ....

Box 1.5 contains some prompt questions, compiled from the support sheets of various teachers, about evaluation.

Box 1.5 *Evaluation: prompt questions*

- Was this test fair?
- How accurate were the observations or measurements you made? (To the nearest mile or the nearest millimetre?) Were they inaccurate because of using the wrong instrument, or because of making a mistake in using it? Could you make it more accurate? How?
- Were your results reliable? Did you need to repeat any results? Why?
- Did you have enough results? Make suggestions about extra results that would have been useful to collect.
- Did you have any odd results? What do you think is the reason for these? Did you need to repeat any results? Why? Say whether you think the pattern of your results is realistic.
- Use your data to test the relationship between the variables. Do your results enable you to find out about different values you haven't measured? (Interpolate and extrapolate.)
- Do these answers make sense? What does this tell you about the quality of your evidence?
- Comment on whether you think your results are good enough to support your conclusion.
- Do you think that the method you used was a good one for testing the prediction you made?
- What other tests could you have done?
- Can you suggest improvements to your method that would make your results more reliable?
- Is there anything else to do with this subject or problem you could investigate that would add to your understanding of it?

Other resources

The current curriculum in England and Wales reflects recent thinking about how students might be expected to carry out scientific enquiries. We have relied heavily on several projects that have greatly influenced the development of the curriculum; indeed, several of the authors of chapters in this book have worked on these projects. The publications from these projects form a core of useful resources that can help any school to develop and deliver its own scheme of work.

◆ *Projects and publications*

- *Science Investigations* is a resource in three packs, designed to improve the investigations students carry out by targeting the area where research shows most are currently failing: handling evidence. It is a development of work on a number of projects by the Exploration of Science Team at Durham University. This is referred to as Collins Science Investigations Pack (1, 2 or 3) in the text and tables of Chapter 2.
  Science Investigations (Pack 1: 1997; *Pack 2*: 1998; *Pack 3*: 1999). Series editors Gott, R. & Foulds, K. Collins Educational, London.

   Building Success in Sc1: Science Investigations for GCSE is a versatile resource that can be used with any scheme of work. It provides a bank of real data on a CD-ROM and is supported by a copiable teacher's guide.
  Building Success in Sc1: Science Investigations for GCSE (2002) Gott, R. & Duggan, S. Folens, Dunstable, Beds.

- AKSIS: The ASE and King's Science Investigations in Schools Project by King's College, London and the Association for Science Education. This collaborative project identified aspects of teaching and learning investigations, where teachers and students wanted support. The following publications were developed in response to these needs:
  Investigations: Getting to Grips with Graphs (1999)
   Investigations: Developing Understanding (2000)
  Investigations: Targeted Learning (2000)
  All by Goldsworthy, A., Watson, R. & Wood-Robinson, V. ASE, College Lane, Hatfield, Herts AL10 9AA.

  The Science Investigator computer program is included in the *Developing Understanding* pack. The original version was published by Essex Advisory and Inspection Services.

- Cognitive Acceleration through Science Education (CASE) Project. The materials are published as *Thinking Science* by Adey, P., Shayer, M. & Yates, C. (3rd edn 2001) Nelson Thornes, Walton on Thames, Surrey.
  To use CASE to full effect requires it to be followed as an enrichment course within the science curriculum, according to the psychological and philosophical underpinning. However, many teachers have found individual units from the materials useful for teaching ideas of investigations.

• Focus Educational, PO Box 52, Truro, Cornwall TR1 1EZ
http://www.focuseducational.com/main.htm

• CONCISE Project (Concept Cartoons in Science
Education) uses cartoons to challenge students' and
teachers' thinking about scientific concepts. They can be
used as a starting point for enquiries or as a focus for
evaluating alternative ideas. *Concept Cartoons: promoting
teaching, learning and assessment in science* (2000) Naylor, S.
& Keogh, B. Millgate House Publishers, Sandbach,
Cheshire.

• PRI (Pupils Research Initiative) resource, published by
Collins, London (see Chapter 6, page 114, for more details).

• SATIS (Science and Technology in Science) project units.
ASE Booksales, College Lane, Hatfield, Herts AL10 9AA.

References

DfES (2001) *Key Stage 3 National Pilot: Science: Tutor's
materials.* Department for Education and Skills, London.
(By the time this book is published the revised materials for
the full KS3 Strategy will have been published, containing
similar material.)

DfES (1999) *Science: The National Curriculum for England.*
Department for Education and Skills, HMSO, London.

Gott, R. & Duggan, S. (1995) *Investigative Work in the Science
Curriculum*. Open University Press, Buckingham.

Gott, R. & Foulds, K. (1997, 1998, 1999) *Science
Investigation Packs 1, 2 & 3*. Collins Educational, London.

Her Majesty's Chief Inspector of Schools (2000) *Progress in
Key Stage 3*. OFSTED, London.

Millar, R., Lubben, F., Gott., R. & Duggan, S. (1994)
Investigating in the school science laboratory: conceptual
and procedural knowledge and their influence on
performance in *Research Papers in Education*, **9** (2), pp.
207–248.

Osborne, R.J. & Freyburg, P.S. (1985) *Learning in Science:
the Implications of Children's Science*. Heinemann, Auckland.

QCA (2000) *Science: a Scheme of Work for Key Stage 3*.
Qualifications and Curriculum Authority, London.

2 | *Investigations: Collecting and using evidence*

Ros Roberts and Richard Gott

School science is about the ideas that form the body of what we know as biology, chemistry and physics, together with the ways in which scientists set about collecting and using evidence. Scientists use evidence either to create 'new' knowledge or to solve problems. Collecting and using evidence can be seen as comprising the use of a set of 'tools' to do with defining variables, making measurements, using tables and graphs and so on, and the way in which these tools are put together to design a complete experiment (the so-called 'scientific method' or way of investigating).

2.1 Learning about evidence: a toolkit of ideas

In this chapter we shall concentrate on the ideas that underpin the toolkit and that are a necessary but insufficient part of any such 'scientific method'. These ideas we have called 'concepts of evidence' (Gott & Duggan, 1995).

Learning about evidence matters. You don't need to look far for issues where ideas about evidence are central to questioning what is going on. The recent example of foot and mouth disease is a case in point. The debate in the media and the concerns of the public seldom touched on the central ideas about the *evidence* for how the virus was spread. This should have been the key to making decisions on how the disease could be combated. Targeting questions to find out what was known about the spread might have reduced the widespread public reaction, which was informed by little more than old wives' tales and almost medieval practices. To get to the endpoint of critical engagement with issues, we suggest that it is not enough that students are able to *do* investigations. They must also come to an *understanding of why* they are done that way and what weight can be placed on the evidence that results, and what might be wrong or inconclusive about evidence they are examining. It follows, then, that we are looking for progress in our students, not simply of performance but also of understanding the ideas that underpin that performance.

When we think about progression in investigations, what is it that is to progress? As discussed in Chapter 1, this could be the level of *manual dexterity*, the sophistication of the *context* or of the *explanatory ideas* that underlie the experiment, or the *underlying ideas* related to design, measurement and analysis. We believe that it is the last of these that should be the driving force. We will concern ourselves with fundamental understanding rather than a ritual approach to experimentation that accords value to mantras such as 'a line graph will get you level *n*' or 'always do three repeats'. To develop this understanding we need to teach the ideas from the toolkit.

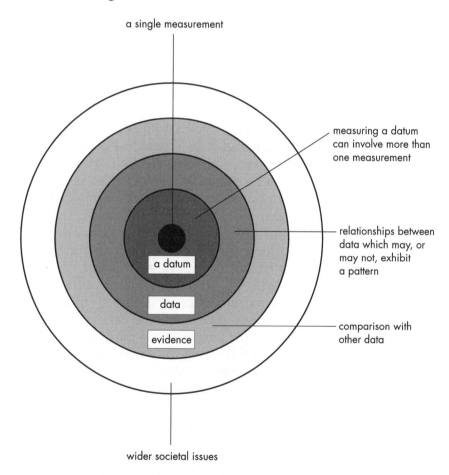

Figure 2.1
Data and evidence (from http://www.dur. ac.uk/richard. gott/Evidence/ cofev.htm).

The ideas in the toolkit come in different 'sizes' so to speak. To begin with, any experiment will require something (a datum) to be measured in some valid and reliable way (Figure 2.1). Individual measurements, taken together, will give the data from that experiment. The experiment has to be designed, again in a valid and reliable fashion. The best check on reliability is triangulation – either by checking what results come from other people doing the same experiment, or by conducting a different experiment designed to test the same thing. Finally, at the 'large end', comes the judgement stage, in which questions like the following are asked. Is the evidence good enough? Is it biased in any way – through the way the research was conducted or funded, for instance?

This chapter is principally about measurement, design and reporting. Design is discussed further in Chapter 3 and evaluation is dealt with in Chapter 5.

◆ *Teaching and assessing the ideas in this chapter*

People often assume that ideas to do with experimental work are encountered only in practical work. It is not as simple as that, however. The ideas we are talking about here, as with all ideas, can be put to use, and taught, in a variety of contexts. They can be taught and assessed using both practical and non-practical work. The important point is that they need to be *explicitly taught* as well as being assessed. Some students will pick up these ideas in the course of studying the traditional content of science, but many will not.

This brings us to the particularly difficult question of choosing a route when dealing with ideas about evidence. Most schemes of work will be driven by the substantive concepts of the curriculum – particle theory, speed and velocity, reproduction, or whatever. But the ideas that we are talking about here do not follow the same path. For instance, much of biology lends itself to introducing and reinforcing ideas about sampling, control experiments and tabular forms of data display. Physics, on the other hand, favours laboratory-type experiments with careful controls, continuous variables and line graphs. The danger, then, is that the curriculum of ideas about evidence becomes a very fragmented and non-linear affair, subject to whatever fits in with the topic sequence of the curriculum. There is no escape from this conundrum. All we can suggest is that teachers have a list of the ideas that need to be covered, with some notion of progression, and map these on to their own schemes of work as best they can.

We shall suggest a route through the ideas in the following text, but it is somewhat artificial in a brief chapter. This is because the ideas are all interrelated and within each area there is a progression, some ideas being harder than others, so that each area can be revisited several times as students progress through secondary schooling.

A valuable starting point is to set the scene, so that students understand why good data are important and to give them an understanding of the structure of an investigation so that they can 'have a go'. This involves their knowing some of the ideas about variable structure and design. The 'reporting' ideas run parallel to the ideas about how evidence is obtained. They are presented here as a separate section but teachers will want to select ideas appropriate to the work being done.

In this chapter, the scene is set with a case study about the threat of pollution from burning a new fuel in a cement works. Here it is important to be able to judge data and how it was generated. This scenario is then used as a structure to illustrate the different areas of the chapter.

♦ *Getting started*

Case study 1: That chimney

The headlines on page 22 are similar to those in a local paper in a dale in north-east England. They refer to a cement works situated in a valley in an area of outstanding natural beauty. A few years ago the management of the factory decided to change the composition of the fuel being used. They started to mix recycled liquid fuel (made up of a variety of waste products) with the original fuel. The factory responded to public concerns about pollution from the chimney and agreed to monitor the resultant emissions for a trial period. Local residents were concerned about the new emissions from the chimney and their effect on health, the water supply and the neighbouring farmland. Groups of residents challenged the findings of the experts brought in to gather data for the company and burning of the new fuel mixture was eventually stopped. A couple of the locals were practising scientists; others involved had specific local knowledge and became sufficiently confident to realise that they could ask sensible questions about how the data was collected and was being used. (Research into this has been reported in Tytler, Duggan & Gott (2001a, b).)

'New fuel used in factory: fears of increased pollution'

'Locals dispute experts'

'We never used to get covered in dust'

'They only sample up the dale. What did you expect to find up there?'

'More kids getting asthma – it's that chimney'

'Villages downwind get the worst of it'

'Experts only take readings that suit their case'

'"The sheep are suffering," says farmer.'

Our research has uncovered the issues that influence the locals' action – bias, politics, etc. Also, and key to this, were issues to do with sampling, instrument sensitivity and so on. That's why teaching about these ideas matters.

◆ *The ideas in the toolkit*

We have produced a list of ideas that are important in collecting, using and interpreting evidence. They can be found at http://www.dur.ac.uk/richard.gott/Evidence/cofev.htm. A selection relevant to this chapter is shown in Table 2.1 and exemplified from the case study.

Table 2.1 *The knowledge base about evidence*

Idea	Definition
Validity	This is the key to everything. If an experiment is not valid, then any data resulting from it are of very limited, if any, use. What does the term mean? It depends on measuring the appropriate thing to obtain data relevant to the issue in question. For instance, in the case study if samples of pollution are taken upwind of the chimney then no amount of care with measurements to get reliable results will be any good. The validity will be compromised if the instruments the scientists are using measure for heavy metal pollution when the real problem is dioxins
Reliability	This refers to the extent to which data can be trusted and is usually linked, therefore, to issues of measurement. An instrument that gives different readings for the same level of pollution is an example here
Variables	The independent (input) and dependent (outcome) variables define an experiment. The type of variable (categoric or continuous) limits the use that can be made of the data and determines the sort of table and graph that is appropriate. In this case the variables are 'type of fuel' (independent variable: categoric) and 'level of pollution' (dependent variable: continuous)
Choice of instruments	How 'good' does an instrument need to be to answer the question? This can often depend on cost – good instruments cost more! In the cement case, the measurements are difficult and expensive to make
Instrument sensitivity	Can the instrument detect whatever it is we are trying to measure at all? In the cement works example, the pollution levels the scientists were measuring were very small, close to the sensitivity of the instruments; hence they were in the 'noise' zone and prone to unreliability
Unavoidable variation	Some variables simply cannot be controlled however hard we try. In the case study, there will be variation in the make-up of the fuels from one batch to another, variation in the emissions when the furnace is running at different temperatures, plus untold numbers of environmental variables (outside temperature, wind speed and direction, humidity, etc.); these will result in variability in the measurements and this has to be taken into account in coming to a judgement
Sampling	Whenever anything is sampled, there is an issue about the representativeness of that sample. A sensible sampling frame is a key element in the cement case study. There is no point, as we mentioned above, in sampling upwind of the chimney, or directly underneath it for that matter. The sample has to cover the key area where the plume of smoke 'lands'
Analysis	The analysis and reporting of the whole investigation must link to the design and variables

Not all of the ideas on the website are appropriate to secondary school level. But some quite definitely are. We are now going to select some of these ideas and see how they might be developed in a series of lessons and also how they might be assessed. We shall draw heavily on Collins Science Investigations packs. See 'Other resources' in Chapter 1 and References to this chapter.

2.2 Measurement

We shall begin with measurement, which is at the heart of much science. When students carry out their own investigation, how often do you hear the plaintive cry, 'my results are wrong', followed by a dash to copy down some 'right' ones? Underlying this reaction is the assumption that science deals in 'right' numbers and so the students themselves must be personally at fault. Everything is attributed to human error.

This assumption needs to be challenged over and over again. It is central to society's failure to approach scientific evidence for what it is – a more or less uncertain affair that can be relied on to some extent, but to an extent difficult to judge.

◆ *Previous knowledge and experience*

Students will be familiar with repeated readings that differ. If you ask them the cause of the variation, they usually blame themselves. They are not aware of the ideas that determine the selection of instruments and the cause of variation in repeated readings. They often rely on a predetermined number of repeated readings. Students recognise thermometers and balances as instruments. Most would not realise that the ideas relevant to instruments also apply to biological measures such as the use of indicator species.

◆ *Teaching suggestions*

The ideas we shall cover are listed in Table 2.2, together with other materials that might be useful.

Table 2.2 *Ideas of measurement*

Ideas from the toolkit considered in this section	Further examples of teaching strategies that target this area
Selection of instrument	Collins *Science Investigations* Pack 2, Unit 2 'Choosing your instrument'
Validity of measurements	Collins *Science Investigations* Pack 2, Unit 2 'How do I measure that?'
Range and interval	Collins *Science Investigations* Pack 2, Unit 2 'Range and interval' Folens' *Building Success in Sc1*
Repeats	Collins *Science Investigations* Pack 2, Unit 2 'Enough data?' and 'Taking the measurement' Folens' *Building Success in Sc1*
Causes of variation in measurements	Collins *Science Investigations* Pack 2, Unit 2 'Always the same?' Folens' *Building Success in Sc1*
Biological indicators	Background information and ideas that can be adapted in Cadogan & Best *Environment and Ecology* and Dowdeswell *Ecology Principles and Practice*
Estimating	Collins *Science Investigations* Pack 2, Unit 2 'Estimating'

Let us start with a lab-based investigation which links to the cement works issue. Any factory requiring energy from a fuel will be concerned at the heat output of the fuel and any pollution problems associated with its burning. The ideas suggested in the next section are not dissimilar from those of the dispute in the case study. A sensible link can be made to it by way of introduction.

The ideas about measurement are first addressed in the more familiar lab-based context. The subsequent application of the ideas in relation to fieldwork introduces a progression in the ideas from the toolkit.

◆ *Measuring how good a fuel is*

1. You could begin with a class brainstorming activity to think about what might be considered 'good' and record the suggestions on the board. Students will come up with different characteristics or variables: the heat output, the power or the cleanliness. They could discuss which of these they think is the best measure of 'good' and vote on it, giving reasons for their decisions.

2. Students then need to decide the best way of measuring this variable. Again, a class discussion or a poster presentation (with notes on the diagram showing why it is like it is) is useful for this. Is it good enough just to say that a fuel is dirty, or makes things hot? Students should recognise that the most useful data are quantifiable. They are most powerful (in terms of the evidence obtained from them) if they can be presented in numerical form. These ideas can be illustrated with reference to measures of heat output; the change in temperature in some water would give a good measure.

3. How much water should be heated by the fuel? The idea of trialling to select the value of variables can be taught here by using a teacher demonstration. Figure 2.3 shows the apparatus set-up.

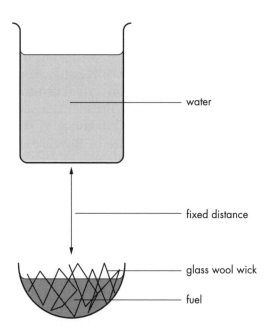

Figure 2.3
Apparatus set-up.

You can show that if you use too little water it will boil before the fuel has finished burning (a common mistake made by students that can be related to the validity of measurements), whereas using too much gives you a very small change in temperature that is difficult to measure accurately. Students should recognise that the volume is about right when there is a sensible increase that can be measured accurately by the instrumentation to hand.

4. This could lead to a discussion on the choice of the instrument. You could use thermometers with different ranges or different beakers, cylinders or jugs to measure the volume of water to be heated. For instance, it is inappropriate to use a 500 ml beaker to try to measure 100 ml of water accurately; the 100 ml measuring cylinder with smaller divisions is more likely to give an accurate measurement. The sensitivity of the instrument, which refers to the smallest change in the value that the instrument can detect, is also something to be considered.

5. Now is the time to let the students have a go. Get them to record their results on the board or on a class spreadsheet (Figure 2.4).

Figure 2.4
Students' results.

The inescapable variation in the measurements will soon be obvious! Some of it will be down to human error, but not all of it by any means. It might be useful to refer to the variation as being made up of (human) error, and (unavoidable) uncertainty or variation. The idea of variation must be distinguished from 'variables' to avoid a common confusion (see also Chapter 3). The variation can be summarised by finding the mean and the range. A scattergraph of the repeated readings (Figure 2.5) could be plotted either by hand or from the spreadsheet. This is discussed further in Chapter 5.

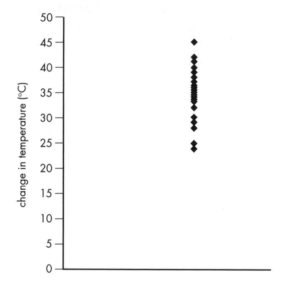

Figure 2.5
Scattergraph of points around a mean.

6. This is a good time to raise the meaning of the average. Students tend to think of this value as the *right* answer. So where does it lie? It is probably somewhere near the middle, but we can't be certain. This uncertainty can be recorded on a graph as a block or line.
7. Why aren't the readings all the same? Students' first reaction may well be to blame themselves (they didn't read the thermometer correctly, or hadn't left it long enough to get to temperature). This 'human error' may well be a contributory factor but is by no means the only thing to cause variation. To be able to get exactly the same result for every repeated reading would require everything to be managed *exactly* the same each time: the volume of water and fuel, the distance of the fuel from the water, air movements by draughts, the heat capacity of the beaker, etc. There is also the consideration of the possible

uncertainty that might be inherent in any measuring instrument. You might ask is 10 ml in the cylinder really 10 ml? Is the thermometer calibrated? Students could be asked to list on a poster all the possible sources of uncertainty. Which of these were outside their control? Which could have been avoided? These ideas could be extended into a 'measurer's manual' of things to think about to make good measurements and this could be displayed in the lab or put as a reference in their books.

8. When students have recognised that there is unavoidable uncertainty in any measurement, it is reasonable to consider just how many repeated measurements are required. The answer to this is definitely not a predetermined, set number! Students should look at their table of repeated readings (in 5 above). As each reading is plotted on to the scattergraph they should be considering whether there is a point where increasing the number of repeats adds little more to the overall picture of the variation (Figure 2.6). How many repeats are needed to 'capture' the unavoidable inherent variation?

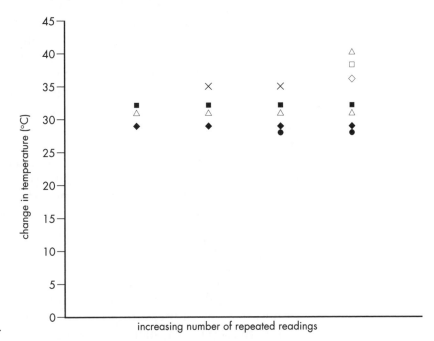

Figure 2.6
Plot of increasing numbers of repeated readings.

9. Students could be given data from repeated readings of different events, some with very little variation, others with far more. How many repeated readings would they do for each event?

◆ *Other measurement ideas from the toolkit directly relevant to this context*

1. The instrument used by students is often determined by availability. However, if there is a choice, they will need to understand how to make a decision. In the fuels investigation several instruments need to be used – to measure the temperature, the volume of fuel and water and the distance of the fuel from the beaker. You could provide students with a choice of beakers and measuring cylinders with different scales to choose from. Since there is inherent uncertainty in all instruments, students should be taught to choose those that have small graduations and are nearest to their full scale deflection to minimise the uncertainty in every reading they take.

2. You can introduce the ideas of accuracy and precision using thermometers and balances. Precision refers to how closely repeated readings are to the same value, and accuracy to how close they are to the 'true' reading. Most thermometers are precise. One will give the same reading each time it is used in something of the same temperature. However, school thermometers are usually fairly inaccurate: they don't give the 'true' reading. Students can investigate this by seeing the reading on their thermometer in boiling water. It is not unusual for school thermometers to be several degrees out both from the 'true' reading and from each other! The more expensive thermometers tend to be more accurate.

   Similarly, if a 100 g mass is put on a selection of balances, not only is it likely that this will show up inaccuracy in the balances but often if the mass is removed and then replaced you might then get a different reading – the instrument is imprecise. Students need to be taught that this matters; you can't put much trust in your readings if they are not reliable.

3. This leads nicely on to the idea of triangulation. Instruments should be checked against another instrument or another way of measuring the same thing.

4. Some investigations, such as the fuels task, involve lots of different measurements. Students could list all the things that need to be measured: the distance between fuel and water, the volume of water, etc. (It is worth emphasising that measurements are needed to standardise control

variables as well as to measure outcomes, as discussed further in Chapter 3.) In the knowledge that every reading can contribute to the uncertainty of the data collected, students could decide how great the uncertainty might be and whether anything could be done to reduce it. Titrations are another excellent context in which to consider this idea, where the independent variable also needs to be measured.

5. In the case study on page 21, the sensitivity of the instruments was crucial. The measurements were very close to the limits of the smallest amount that could be detected by the instrument. Students could determine the sensitivity of various instruments around the lab and decide whether these would be any good for measuring different values.

6. If certain fixed points are known, such as the temperatures of melting ice and boiling water, then instruments can be calibrated. Students may be familiar with this from their work on the expansion of liquids with temperature. A good activity is to get the students to make their own thermometer that is as precise and accurate as possible and to use it to measure something of unknown temperature.

♦ *Measuring pollution: an example from biology*

Applying the same measurement ideas to a more complex biological situation provides an opportunity for progression in the ideas of measurement.

1. Pollution can be measured by sophisticated instruments that detect and quantify specific pollutants (Geiger counters, pH meters, gas chromatographs, etc.). In a school situation, however, it is often necessary, and sensible, to use more indirect methods. Pollution has, by definition, a harmful impact on living things. A class brainstorm might come up with such things as the effects on the number of organisms surviving, the species diversity (since some species may be killed off), the size of animals and plants or physiological responses such as respiratory conditions in humans. Could any of these be used to measure the pollution? Students could discuss what it is they would need to know about these effects before they could use the information as a basis for making measurements.

It is worth pointing out that such a way of measuring pollution is a measuring instrument just as surely as is a measuring cylinder. The key criterion for a measuring instrument, of course, is that you get the same answer each time it is used (in the same situation).

2. The use of biological indicator species to measure both air and water pollution is widespread. For instance, lichens are particularly sensitive to long-term effects of the levels of sulphur dioxide in the air. In clean air there are more types of lichen present (increased diversity) and the types present tend to be branched and bushy forms rather than flat encrusting forms. Thus the relationship between size or diversity of lichens (or both) and levels of pollution forms the basis of the measuring instrument. Students need to consider the limitations of using biological indicators as measuring instruments, however, since the relationship is itself affected by other variables.

How *reliable* does this make biological indicators as measuring instruments? In other words, do they give readings that you trust? If they are affected by other variables then these will need to be controlled if the organisms are to be reliable instruments. The other potentially significant variables can be shown in a *circle of variables* (Gott *et al.*, 1997) (Figure 2.7), which is discussed in the Design section 2.3.

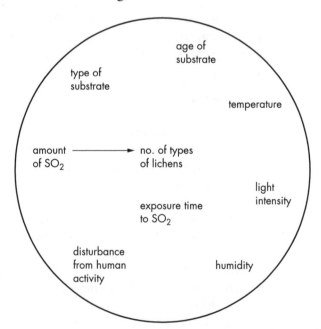

Figure 2.7
Circle of variables.

3. How *valid* are they as measures of pollution? For instance, could they be affected by other things and give false positives, or fail to be affected and give false negatives? Students could research investigations that have been conducted to establish the relationship between pollution and the indicator (for example those shown on the Natural History Museum website).

♦ *Other measurement ideas from the toolkit directly relevant to this context*

1. How good are the measurements taken when using a quadrat? How good is the estimate of, for instance, percentage cover or abundance on a frequency scale? Students could consider the reliability of these estimates, referring to accuracy, precision and sensitivity.
2. What measuring instruments are available for measuring the pH of the rain water? Which are most suitable, and why?
3. Set a task such as: scientists want to investigate whether the pH of rain water is always the same at a particular site. What should they consider when deciding how long to leave the rain-collecting apparatus out for and how frequently to measure the pH?
4. Biological indicators are often more valid measures of pollution than one-off chemical-based measures. This is because their characteristics reflect the conditions at the site over a period of time rather than the 'snapshots' of chemical conditions at the intervals of sampling. Which biological indicators can be used to measure sudden changes in pollution levels? (The canary predecessor to the Davy lamp would be an historical example of such a biological indicator.)

♦ *Assessment (Table 2.3)*

Table 2.3 *Assessment of ideas about measurement*

Ideas from the toolkit assessed	Examples of assessment activities that target this idea
Choice of instrument Range and interval of independent variable Identifying source of variation in readings Dealing with anomalous readings	Collins *Science Investigations* Pack 2 'Making measurements', test and answers

2.3 Design

♦ *Previous knowledge and experience*

Students will have had varying previous experiences. Most, from junior school, will know how to do a fair test 'by keeping everything the same'. Their understanding of *why* they are doing what they've been taught to do will be very limited. Few will understand the ideas that underpin all types of science investigations. Our research has shown that most students have difficulty in:

- understanding validity – in the limited sense of knowing that some data is rendered worthless if it is measuring the wrong thing
- identifying variables
- understanding variable structure in a fieldwork (i.e. non-manipulative/out of lab) context
- understanding control – controlling for the effect of variables, not just routinely 'keeping everything the same'
- understanding that tables and graphs are not just display tools, but can be an essential element of the experimental plan and an aid to modifying the plan in practice
- holding all the ideas together in designing a practical task.

◆ *Teaching suggestions*

The ideas we shall cover are listed in Table 2.4, together with references to other materials that might be useful.

Table 2.4 *Ideas of design*

Ideas from the toolkit considered in this section	Further examples of teaching activities that target this idea
Tables as organisers	Collins *Science Investigations* Pack 1, Unit 2 'Personal organiser' and 'Top table'
Selection of variables	Collins *Science Investigations* Pack 1, Unit 1 'Prove it!'; Pack 2, Unit 1 'So many variables'
Fair testing and validity	Collins *Science Investigations* Pack 1, Unit 3 'Is it fair?'
Values of variables	Collins *Science Investigations* Pack 1, Unit 3 'Is it fair?' and 'How much?'
Types of variables	Collins *Science Investigations* Pack 1, Unit 1 'So what's different?'
Defining the question	Collins *Science Investigations* Pack 2, Unit 1 'Long investigations'
Selection of variables	Collins *Science Investigations* Pack 2, Unit 1 'So many variables'
Selecting the value of variables	Collins *Science Investigations* Pack 2, Unit 1 'Designing investigations'
Control in fieldwork	Collins *Science Investigations* Pack 2, Unit 1 'Designing investigations'
Representative sample of the DV	Collins *Science Investigations* Pack 2, Unit 1 'Representative samples'
Sample size	Collins *Science Investigations* Pack 2, Unit 1 'Random samples'
Correlation and causation	Collins *Science Investigations* Pack 3, Unit 1 'Cause and effect'

Having considered the problems of measurements, students can be given a complete task to do and see if they can use the ideas. Here two different questions are considered that illustrate the different emphasis placed on the ideas in lab-based and fieldwork contexts. In each case a number of different activities are suggested. The number of ideas required to be able to carry out the whole investigation unaided is extensive. This is why students find whole investigations difficult. At the start of teaching you will have to decide which ideas are to be taught in the context provided. You will consequently have to decide which decisions you expect the students to make and which decisions you will make for them. A list of key ideas on the board with frequent exhortation to 'think about this as you go along' is helpful in keeping their ideas in focus.

◆ *Which is the best fuel for boiling water?*

You can use both practical and non-practical activities to teach the ideas required to investigate this question. It can be linked to the case study on page 21, to the ideas about measurement in the previous section and to reporting in the next section. Further activities for defining questions and identifying variables are offered in Chapter 3.

! *Do a risk assessment.*

Ethanol, methanol and butanol are suitable fuels to use.

Care with fuels and flames! Use goggles and take appropriate precautions.

1. You need to define the question: 'what do we mean by "best"?' This time the best might be defined as the quickest. What variables are involved? In this lab-based investigation the independent variable is changed by the investigators: they burn each fuel in turn. The dependent variable will be 'power output'. Identification of independent and dependent variables takes practice. We can draw a circle of variables as a useful way of presenting this and helping with the organisation of the experiment (Figure 2.8). Students will be familiar with the idea of control variables as those whose values have to be kept the same to ensure a fair test.

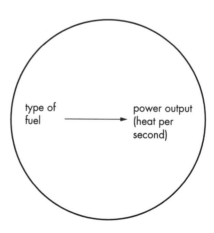

type of fuel ⟶ power output (heat per second)

Figure 2.8
Simple circle with key variables.

2. We can then draw another variable circle that now includes the 'measurement' element of the task. The new dependent variable will be something that can be used as a measure of power output – the time to bring to the boil a fixed quantity of water perhaps (Figure 2.9).

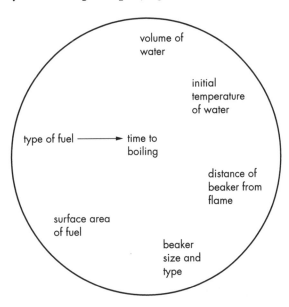

volume of water

initial temperature of water

type of fuel ⟶ time to boiling

distance of beaker from flame

surface area of fuel

beaker size and type

Figure 2.9
Circle of variables for fuel/boiling.

3. Students need to be able to understand that the *values* of categoric variables are expressed in words, and the *values* of continuous variables as numbers. They could annotate the circle of variables they produce, indicating whether the variables are categoric or continuous.

4. Tables (and graphs) should be seen as proactive, not just as display tools for tidying up and recording data that has been jotted down haphazardly or as an afterthought after the practical part of an investigation has been done. It is far more useful to use the table as an essential element of the experimental plan – in other words, get students to construct a table *before they begin*. Tables are also, of course, a report of the actual data. So tables have a dual purpose:

   • as an organiser for the investigation
   • as a way of presenting data in a report.

   You can focus at first on the use of a table as an organiser for an experiment. The same table can be used in a report or if necessary redrawn, for example, using a computer. The structure of a table is linked to the variable structure as shown in Figure 2.10.

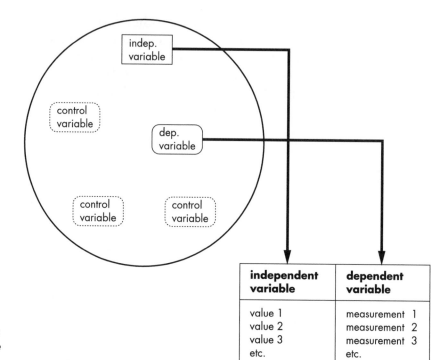

Figure 2.10
*Circle of variables
related to structure
of table.*

independent variable	dependent variable
value 1	measurement 1
value 2	measurement 2
value 3	measurement 3
etc.	etc.

5. Before students can 'have a go' at the investigation themselves, they need to consider the *values* of the variables. In the case of categoric variables, the values are determined by the types available (i.e. the types of fuel provided). However, the values of the continuous control variables (such as how much fuel to use, how much water to put in the beaker and how far the fuel should be away from the beaker) can only be determined by trialling.

6. When designing the investigation it is impossible to predict the number of repeats that will be necessary to determine which fuel is best. The decision can be made only *after* any variation in the repeated readings is found empirically. The treatment of repeated readings, and making such a decision, is discussed in teaching suggestion 4 on page 45. Students following the same procedure could gather class data for the variation in readings from all three fuels. Is it possible to distinguish between the fuels after only three repeats? How many readings are needed? The answer to this question will depend on the data obtained and cannot be known in advance.

Alternative strategies for identifying variables and organising the design of an investigation are included in Chapter 3.

◆ *Does pollution have an effect on living things?*

The concern of the residents in the case study on page 21 was for possible pollution effects. The suggested activities can all be linked to this. The same ideas are as important in a fieldwork context as they are in the lab, but with a different emphasis. Many students find the 'transition' from lab to fieldwork a difficult progression. They do not perceive that the same ideas underpin the two contexts. A range of activities is suggested to help them. However, not all schools are able to work with students off-site so suggestions are also made for activities that can be carried out by students in the school grounds and in the classroom.

> **!** *A risk assessment must be performed and safety procedures specified by the school must be followed when taking students out of the classroom or off-site.*

1. Many students have difficulty turning a question into an investigation with clearly defined variables. You could ask them to work in small groups to suggest how to investigate the question: 'does pollution have an effect on living things?' Students need to be able to identify the independent variable. Younger students may come up with something quite general such as clean or dirty air, or the presence or absence of pollution. Those with more understanding of the causes of pollution may specify the type of pollutant, such as sewage, factory run-off, heavy metals, acid rain or sulphur dioxide emissions. How this might be measured is discussed in 'Measuring pollution' on page 31.

2. A circle of variables could be used to identify other variables that might affect the dependent variable (Figure 2.11). How might their effects be controlled? Control in field investigations cannot be achieved by manipulating the value of the control variables. So the possible effects of the other variables need to be reduced by taking measurements at sites where significant dependent variables are already similar. Students need to consider the question: 'which variables would need to be measured to ensure the comparison was fair?'

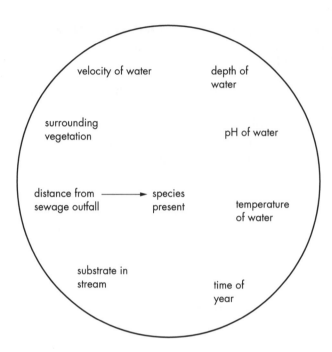

Figure 2.11
*Circle of variables:
pollution.*

3. Is one reading of the dependent variable enough? As
   before, the inherent uncertainty in the measurement
   requires that repeated readings are taken to 'capture' the
   variation. In the case study on page 21 the scientists'
   evidence was challenged on the grounds that the
   measurements were taken from only a few sites that didn't
   represent the variation in the whole area. There are several
   ways to ensure that the sample is representative. It could be
   a large random sample. Or it could be stratified, taking
   readings in appropriate proportions from known sections of
   the range, to ensure the known variation is 'captured' in the
   sample. A simple exercise to get across the idea of the need
   for a sampling frame is to model the situation by playing a
   game like 'battleships' (Table 2.5). Students need to work
   in groups with one person having sheet B and the others
   having copies of sheet A. Players with sheet A have to
   devise strategies so that they can 'capture' the distribution
   of the pollution from the factory with as few readings as
   possible. Players with sheet A call out the grid references of
   the site where they are going to take a measurement. Player
   B then tells them the value of pollution at those sites to
   record as data.

Table 2.5 *Sampling frame*

Grid A

Key:

- = river;
- = housing;
- = nature conservation sites;
- = woodlands;
- = factory.

	1	2	3	4	5	6	7	8	9	10	11	12	13	14
N										river				↑N
M			woodlands				nature conservation			river		woodlands		
L	woodlands		housing						river		woodlands	housing	housing	housing
K									river		housing	housing		
J	woodlands			nature conservation	woodlands		river	river						nature conservation
I			woodlands			river				nature conservation		woodlands		
H		nature conservation	river	river	river		factory		woodlands					
G	river	river										woodlands		
F					woodlands				woodlands					
E			woodlands	housing	housing	housing		woodlands						
D			housing	housing			woodlands				woodlands			
C		woodlands									housing			
B	woodlands													
A														

Grid B

	1	2	3	4	5	6	7	8	9	10	11	12	13	14
N	0.2	0.4	0.1	0.2	0.1	0.3	0.9	1.1	1.2	0.9	1.1	1.7	2.2	1.2
M	0.3	0.2	0.4	0.1	0.2	0.1	0.3	1.2	5.2	8.5	9.1	8.1	1.9	1.8
L	0.3	0.2	0.1	0.1	0.1	0.3	0.2	1.0	5.6	7.3	9.0	8.7	2.5	2.1
K	0.3	0.2	1.5	1.4	0.2	1.1	1.5	0.8	7.3	9.8	3.9	3.6	3.2	2.9
J	0.2	0.4	0.1	0.2	0.1	0.3	0.3	0.5	0.9	3.8	1.2	0.1	0.3	0.1
I	0.1	0.1	0.4	0.2	0.1	0.2	0.3	0.2	0.4	0.2	0.1	0.2	0.1	0.3
H	0.1	0.1	0.2	0.2	0.1	0.1	factory	0.2	0.2	0.1	0.1	0.4	0.2	0.1
G	0.2	0.1	0.1	0.4	0.2	0.1	0.2	0.1	0.3	0.2	0.1	0.3	0.1	0.0
F	0.2	0.4	0.1	0.1	0.2	0.2	0.1	0.2	0.1	0.1	0.4	0.2	0.1	0.2
E	0.1	0.5	0.3	0.2	0.8	0.4	0.3	0.2	0.2	0.4	0.1	0.3	0.1	0.2
D	0.1	0.1	0.2	0.4	0.1	0.1	0.5	0.2	0.2	0.1	1.6	0.5	0.4	0.2
C	0.3	0.2	0.1	0.0	0.1	0.0	0.2	0.8	0.4	1.4	0.0	0.2	0.2	0.2
B	0.5	0.2	0.3	0.2	1.5	1.4	0.2	1.8	0.2	0.2	0.1	0.2	0.1	0.4
A	0.2	0.1	0.2	0.2	0.1	1.2	0.1	0.2	0.4	0.1	0.1	0.1	0.5	0.4

4. How big should the sample be so that it is representative? A simple exercise can be conducted on the school field or garden lawn for this. Students need to be able to distinguish, but not necessarily name, the leaves of different species. You should first define the area in which sampling is to take place, for example the hockey pitch or an area marked out by string. Students place their quadrat frame in the area by using random numbers to determine coordinates. (They could devise a 'random number generator', such as the final 4 digits in telephone numbers, or picking pieces of paper with numbers on out of a 'hat', or could use random number tables or the random number generator on some calculators.) They pick a leaf from each type of plant they find and stick it on their recording sheet, giving it a letter in sequence (Table 2.6). They record the name or code of the species they find in this quadrat. They then 'throw' the quadrat again and pick a leaf of any *new* species they find. The number of new species found is added to the number already found. As the number of quadrats increases, the number of new species decreases. This could be plotted as a graph. A decision can be made as to how many quadrats are likely to 'capture' the species diversity in the area.

Table 2.6 *Quadrat and leaves*

Number of quadrat						
1	**2**	**3**	**4**	**5**	**6**	**etc.**
Species not previously recorded:						
Leaf a	Leaf f	Leaf h	None	Leaf j	Leaf l	
Leaf b	Leaf g	Leaf i		Leaf k		
Leaf c						
Leaf d						
Leaf e						
Running total = 5	Running total = 7	Running total = 9	Running total = 9	Running total = 11	Running total = 12	

If you haven't got access to a grassed area, it can be modelled by a sheet of printed newspaper. The exercise can be conducted using the newspaper sheet and a microscope cover-slip as a quadrat, with letters and punctuation as plant species. It can also be demonstrated on an overhead projector using a printed acetate and a card frame as a quadrat.

5. There are many enhancement opportunities, for example:

   - discuss other sampling frames: line transects, point frames, time sampling, pitfall traps, etc. and when each of these would be appropriate
   - join in surveys carried out by the Royal Society for the Protection of Birds (RSPB), the local wildlife trust and other organisations
   - design and carry out studies to build up longitudinal data by repeating the same measurements in consecutive years, looking for changes over time
   - investigate how species diversity changes with the seasons.

6. Many of the questions that students might raise about the effects of pollution will have been investigated by others. In evaluating reports, students can apply their ideas of experimental design by looking critically at investigations by research scientists or other students (see Chapter 5).

♦ *Assessment (Table 2.7)*

Table 2.7 *Assessment of design*

Ideas from the toolkit assessed	Examples of assessment activities that target this idea
Defining the question, selection of variables, variable type	Collins *Science Investigations* Pack 1 'Making connections', test and answers
Identifying variables, control experiments evaluating the design of investigations, fair testing	Collins *Science Investigations* Pack 1 'Designing investigations', test and answers
Variation, sampling, validity in relation to sample, control variables in fieldwork	Collins *Science Investigations* Pack 2 'Investigating biology', test and answers

2.4 Reporting

◆ *Previous knowledge and experience*

Students will know the mechanics of entering data into tables and graphs, and of writing reports under standardised headings. Again, however, their understanding of *why* they are doing what they've been taught may be very limited. Many students find writing about their work tedious and an 'add on' to their investigation. Few see that good recording is an essential 'organiser' for their work that can guide their thinking and what they decide to do. Often students will:

- have difficulty in constructing tables themselves
- not reorder the data to show patterns
- think that their account should read like a 'recipe' for an illustrative practical rather than an iterative account that demonstrates their understanding of the ideas in the toolkit
- need to be convinced that their report is an essential part of how other people judge their evidence
- not know to how many decimal places to record measurements or calculate derived quantities.

◆ *Teaching suggestions*

Ideas about reporting and recording were introduced in the previous section. This section addresses further ideas (Table 2.8).

Table 2.8 *Ideas about reporting*

Ideas from the toolkit considered in this section	Further examples of teaching activities that target this idea
The importance of good reporting	Collins *Science Investigations* Pack 1, Unit 3 'Telling the story' and 'Be convincing'
Bar charts inc. axes, scale, interval	Collins *Science Investigations* Pack 1, Unit 4 'Bar charts' and 'What do they tell us?' Folens' *Building Success in Sc1*
Ordered data	Collins *Science Investigations* Pack 1, Unit 2 'Using a table'
Frequency histograms	Folens' *Building Success in Sc1*
Using physics and chemistry subject knowledge in an investigation	Collins *Science Investigations* Pack 3, Unit 2 'Chemistry' and 'Physics'

Students need to be taught explicitly how to record and report their work. This is not merely display technique but is integral to collecting and understanding evidence.

◆ *Which fuel is best?*

1. Several ideas of importance to good reporting have been introduced in earlier sections, including the circle of variables, construction of tables and presentation of data. Students need plenty of practice of these in a range of contexts.

2. You can give students some ideas for investigations and ask them to design a table for data collection in each case. This helps them to appreciate that tables are powerful organisers of investigations and should be set out prior to any data collection.

3. You could give students tables (from papers, technical journals, brochures or the web) and ask them to identify the relationships being investigated. What else would they need to know about the investigation before they could decide if they trusted the evidence?

 4. Students need to be able to derive values from the data collected. This is a case for the computer. Ask students to enter their data from their experiment either directly into the computer, or later if none is available in the lab. Let them try to:

   - order the data in different ways to show different patterns
   - calculate averages, and the largest and smallest values (Table 2.9).

Table 2.9 *Fuel data*

Fuel	Time to boil: run					Average	Biggest value	Smallest value
	1	2	3	4	5			
A	3	3	2	3	4	3	4	2
B	4	3	3	4	4	3.6	4	3
C	6	7	6	7	7	6.6	7	6

These can now be converted directly into a graph that will show the average and the range. This, in turn, can form the basis of a discussion of the reliability of the results. Students can consider the question: 'is the spread in the results small enough to be sure that there really is a difference between the fuels?' In our artificial example above, C is almost certainly better, but A and B are not readily distinguishable. This continues the activity of teaching suggestion 6 on page 38, and is discussed in further chapters.

5. Using accounts of other people's investigations is a good way to help students understand the importance of good reporting, where all the decisions are made explicit (Figure 2.13).

   Deciding the number of repeated readings is referred to in 'Measurement' and 'Design' on pages 29 and 38 and in Chapters 3–5.

Data on run 1:

Type of fuel	Rise in temp. (degrees C)
A	39
B	37
C	29

A and B are too close together to be convincing evidence either way. C might be quite different. I will need further runs to check.

Further runs:

Type of fuel	Rise in temp. (degrees C)				Range	
	Run 1	Run 2	Run 3	Average	Biggest	Smallest
A	39	41	36	38.7	41	36
B	37	39	38	38.0	39	37
C	29	28	32	29.7	32	28

This confirms the original data and gives us some confidence that there were no rogue results in the first set. C can be eliminated but still impossible to separate A and B. Further readings are necessary. I will add each reading to a scattergraph to see whether the readings cluster differently.

Figure 2.13 *Extract from a good write-up.*

6. When recording data, how many significant figures need to be shown? Students could consider this both for readings off instruments and for derived units. They could make a note in their measurers' manual. The number of significant figures recorded should reflect the level of accuracy that the data represent.

   This decision is also discussed in 'Measurement' on page 30 and in Chapter 5.

◆ *Does polluted water have an effect on living things?*

1. Recording data in biology often involves the collation of class data in a large table on the board or on a class spreadsheet. Students need to recognise that this is linked to the need for a large sample size in many biological investigations because of the inherent variation in much biological material, discussed in 'Design' on page 39.

2. Students could plan how to record the data generated from the following class experiment. Groups grow cress or radish seeds in good light in plug pots. They are watered with water containing various concentrations of copper sulphate. (In low concentration this acts as a fungicide, but becomes toxic to the plants at higher levels.) Students could decide which dependent variable to measure and to record in a class table. Similarly, data could be collected from duckweed grown in beakers containing various concentrations of washing-up liquid or liquid fertiliser.

3. Some of the dependent variables in the experiments above may be categoric, whereas others may be continuous. How could each be represented graphically? Choice of graphs is discussed in Chapter 3, teaching suggestion 3 on page 58 and in Chapter 5, the teaching suggestions on page 99.

4. Plan a stream or pond survey to investigate the distribution of different plants and animals. Make a circle of variables to identify all the variables that might affect the distribution of different species (see teaching suggestion 2 on page 39). Design a table to record class data.

5. Progression to reading tables that consist of data from several dependent variables takes practice. Students should be encouraged to read tables and identify the separate investigations that contributed data. Sources of material include consumer magazines with data in linked tables and graphs, biology texts with complex figures to summarise ecological data, and daily newspapers. Students could interpret these sources, criticise the display of data for clarity and comment on possible misrepresentation.

♦ *Assessment (Table 2.10)*

Table 2.10 *Assessment of ideas about reporting*

Ideas from the toolkit assessed	Examples of assessment activities that target this idea
Making tables, putting data into tables, reading tables	Collins *Science Investigations* Pack 1 'Organising investigations', test and answers
Reading bar charts, making bar charts	Collins *Science Investigations* Pack 1 'Bar charts', test and answers
Making line graphs, reading line graphs, interpolation and extrapolation, describing relationships, questioning the validity of data shown in line graphs	Collins *Science Investigations* Pack 1 'Line graphs', test and answers
More assessment of drawing line graphs, reading line graphs, interpolation and extrapolation, describing relationships, homing in on 'interesting' parts of the relationship	Collins *Science Investigations* Pack 2 'Relationships', test and answers

References

Cadogan, A. & Best, G. (1992) *Biology Advanced Studies: Environment and Ecology.* Nelson Blackie, Glasgow.

Dowdeswell, W.H. (1984) *Ecology: Principles and Practice.* Heinemann, London.

Gott, R. & Duggan, S. (1995) *Investigative Work in the Science Curriculum.* Open University Press, Buckingham.

Gott, R. & Duggan, S. (2002) *Building Success in Sc1: Science Investigations for GCSE.* Folens, Dunstable, Beds.

Gott, R., Duggan, S. & Roberts, R. Durham website: http://www.dur.ac.uk/richard.gott/Evidence/cofev.htm

Gott, R., Foulds, K., Johnson, P., Jones, M. & Roberts, R. (1997) *Science Investigations 1.* Collins Educational, London.

Gott, R., Foulds, K., Jones, M., Johnson, P. & Roberts, R. (1998) *Science Investigations 2.* Collins Educational, London.

Gott, R., Foulds, K., Roberts, R., Jones, M. & Johnson, P. (1999) *Science Investigations 3.* Collins Educational, London.

 Medical Research Council website: http://www.mrc.ac.uk
RSPB website: http://www.rspb.org.uk/rspb.asp

SAPS (Science and Plants in Schools) Head Office is located at Homerton College, Cambridge and website http://www.saps.plantsci.cam.ac.uk/sapshom.html

Tytler, R., Duggan S. & Gott, R. (2001a) Public participation in an environmental dispute: implications for science education. *Public Understanding of Science* **10**, pp. 343–364.

Tytler, R., Duggan, S. & Gott, R. (2001b) Dimensions of evidence, the public understanding of science and science education. *International Journal of Science Education* **23** (8), pp. 815–832.

Investigations: Planning from questions

Nigel Heslop and Valerie Wood-Robinson

Questioning is the origin of scientific enquiry. Children are forever asking questions. Unfortunately, many secondary school students have lost sight of a link between their own questioning and the questions they are asked to investigate in science, which appear either trivial or esoteric. In the previous chapter, we said that you don't need to look far for issues where ideas about evidence are central to questioning what is going on. We should try to use questions from students' own interests or 'real life' to generate questions that can be investigated by scientific enquiry.

3.1 Asking questions and making plans

In this chapter we are looking at planning scientific enquiries from the starting point of questions. We make suggestions for considering a range of questions. We suggest alternative strategies for teaching the design of experiments to obtain valid and reliable evidence. Students need to be explicitly taught to:

- decide upon questions and how they might be investigated
- use science knowledge and understanding, prior evidence and trials
- design experiments
- ensure the validity of evidence.

This chapter draws heavily on examples from the publications of the AKSIS project *'Developing Understanding'* and *'Getting to Grips with Graphs'* (details are given in the section on 'Other resources' in Chapter 1, pages 15–17).

◆ *Teaching and assessing planning*

Planning is hard for students. It is difficult for those at the beginning of secondary school to coordinate the thinking skills with the equipment and safety demands. They are dropped into an unfamiliar teaching room (a laboratory), they don't know what is in the cupboards, they are intimidated by the language and we terrify them with draconian 'lab rules'. Teachers should nurture the skills that the students have and not push them into planning whole investigations before they have the knowledge, overview and techniques to cope with it. The ideas and skills of finding questions, using knowledge, applying techniques and making evidence valid need to be established first.

This requires explicit teaching. As with other aspects of investigating, this may be done through focusing on a single or small group of teaching points to teach separate skills or ideas of evidence. A whole science investigation is a holistic and very challenging activity, so students need to see the outcomes and full nature of the investigative process before they are able to see where planning fits into the scheme. This suggests the paradox of teaching 'planning' *last*, after all the other aspects of investigation have been introduced and practised – but this paradox exists only if you view the teaching as a linear process. In fact, teaching of investigations is an iterative process, as is performing an investigation, with no beginning or end. With a new class, therefore, you might like to let the students 'have a go' at an investigation from their prior experience of 'planning', then start your focused teaching on 'evaluation'. This will help them to understand the quality of evidence they need to collect. You can then embark on structured teaching of the separate aspects of 'planning'. You will need to revisit component ideas as you embed them in 'the whole picture', to clarify, reinforce and progress them.

◆ *Previous knowledge and experience*

The raft of skills and techniques that students have developed in primary school needs to be recognised and developed within their new environment of more formal laboratory work. Students arrive from primary school familiar with the literacy skills of framing questions. However, they may find it difficult to reframe a question so that it can be answered by a scientific enquiry. Many will have experience of a planning tool such as the 'planning house' or the sticky label system, and may regard this as essential to science rather than one of many alternative aids to organising their thinking. They will be familiar with doing a 'fair test' by 'keeping everything the same' but many will be unclear why they do that. They may try to apply a 'fair test' to every scientific enquiry even when it is not relevant.

They may have learnt the terms 'what we change' and 'what we measure' for independent and dependent variables respectively. However, these can introduce misconceptions, for example in investigations where the values of the independent variable and standardisation of control variables also have to be *measured*. Students will be able to write in different styles for different purposes, but some may have been taught a formulaic style for writing experimental plans that do not express their own ideas. Other items of previous knowledge and difficulties about measurement, design and reporting are listed in Chapter 2.

The suggestions in this chapter should help students to progress in their ability to plan investigations by:

- distinguishing different kinds of enquiries to tackle different scientific questions
- refining their questions to more precise and critical outcomes
- working from categoric variables, through ordered variables to continuously measured quantities
- controlling more variables or considering how to deal with uncontrollable variation
- choosing and using more sophisticated equipment and sensitive instruments
- taking more of their own decisions.

3.2 Deciding questions

We need to encourage students to ask questions. We have to teach them how to define questions in such a way that they can be investigated scientifically. This involves teaching them to decide which questions, both their own and from other sources, are amenable to scientific investigation and by what kind of enquiry. The framing of a question so that it can be investigated includes deciding what evidence to look for. This process implies a hypothesis or prediction of the expected relationship or outcome.

We need to teach students:

- to decide on an appropriate kind of enquiry
- to identify the scientific knowledge relevant to a particular question
- to frame a question that can be investigated
- when it is appropriate to seek data from secondary sources rather than first-hand data.

◆ *Teaching suggestions*

1. To help students identify a range of scientific investigations you could present them with wide-ranging questions on a topic, perhaps on cards, and ask them to suggest further questions. Ask the students to sort the questions into those that would be suitable for scientific enquiry (e.g. 'what do woodlice eat?') and those that would not (e.g. 'are woodlice horrible?'). They then discuss the strategies they would use to answer the 'scientific' questions, and sort the questions accordingly. The following lists of 'kinds of enquiries' will guide students and avoid going into detailed plans:

    - surveys and correlations (pattern seeking)
    - using secondary sources (reference)
    - classification (identification and classification)
    - using and evaluating a technique (technology)
    - control of variables (fair test).

    Students should realise that there are often alternative approaches and so there are no right and wrong answers to this activity. The value of the activity is in the discussion.

    This activity is based on the AKSIS Project *Developing Understanding* Unit 4, 'What kind of question, what kind of enquiry?' which includes sets of questions and a description of eight ways of 'How we can find answers'. The QCA *KS3 Scheme of Work*, Unit 9M and the *KS3 National Pilot for Science*, Unit 4 contain adapted versions.

2. Students need to frame a question in such a way that it can be investigated. The question, or statement arising from it, needs to show what kind of evidence to collect to answer the question. What do we have to observe or measure to find the answer to the question? This is discussed in Chapter 2. If we are asking 'what is the best...?' then what is our criterion of 'best'?, and 'best' for what purpose? (see teaching suggestion 1 on page 36). A question like 'does pollution have an effect on living things?' needs to be operationalised to identify which effect of a range of possible effects is to be investigated (see teaching suggestion 1 on page 39). Give students some questions, to identify whether each question tells you what evidence should be collected or how you might collect evidence. They can then reframe the questions to define the structure of the investigation. The framing of the question will suggest the kind of enquiry, as in the examples in Table 3.1.

Table 3.1 *Reframing questions*

Question	Reframed enquiry question	Kind of enquiry
What makes grass grow well?	How does amount of watering affect the height of grass?	Fair-test investigation
What is the best material for a lab coat?	Which of several materials is least flammable and most resistant to acid spills?	Fair-test investigation with two dependent variables
What happens if you mix baking powder and vinegar?	(Question could stay the same)	Exploration: mix as suggested and observe any events
Does pollution affect living things?	How does the amount of nitrate in a river affect the number of animal species?	Correlation of two independent variables (pattern-seeking)
How can we separate rock salt?	How can we separate rock salt into sand and salt?	Using and evaluating a technique (technological)

You may want students to do this task intuitively as an introduction to the idea of variables, or you may want to reinforce that idea by asking them to identify the key variables as such. Alternatively you may want them to reframe a diffuse question *after* they have identified the key variables; this is the approach adopted in *The Science Investigator* program (see teaching suggestion 3 on page 63).

3. Using the DART strategy of highlighting key words and phrases relevant to a particular focus, you can teach students to analyse a passage of information to identify the relevant scientific knowledge. You could use a piece of fiction or non-fiction text that is not specifically scientific. The analysis can help in deciding which specific questions can be investigated. The example in Box 3.1 opposite is based on text in a cookery book (adapted from *Developing Understanding*).

4. Some questions cannot feasibly be investigated first-hand, because of limitations of scale, time, cost, equipment, expertise, etc. Students have to understand that an enquiry using secondary sources goes through similar thinking processes to a practical enquiry. It is *not* a matter of just looking up the answer to the question, or finding everything they can about a topic.

   To teach students to use secondary sources you could set up a 'scientists' conference' on a single theme or covering several topics. Each group of pupils is a team of scientists investigating a question that they suggest or are given. They should imagine that they have unlimited resources to

Box 3.1 *Yeast mixtures: how to handle yeast (from a Penguin Cookery Book)*

Using yeast is not difficult, providing you remember it is a living plant that can be killed by the wrong treatment and that when it is dead it will not work to lighten dough. Yeast needs food and warmth to make it grow. Its food is sugar and flour. As it grows it produces bubbles of gas called carbon dioxide, and this gas makes the dough light. . . . When the mixture goes in the oven, the yeast is killed and no more gas is produced.

Underline in BLUE the parts that tell you yeast is a living thing, in RED the things that are essential for yeast to work, in BLACK anything that can kill yeast, and in GREEN all the parts that suggest that yeast is not really a plant.
  Now write some questions about yeast that you could investigate scientifically, from the ideas you have underlined.

investigate it first-hand, so they must formulate a clear question, *as if* they had the facilities to investigate it first-hand, recognise valid evidence to address the question of the enquiry, and understand ideas of evidence to know what kinds of data and how much data (or how little) to collect to provide reliable evidence. They then search for evidence from secondary sources (reference books, on CD-ROMs, on databases and the internet) *as if* their team had collected it first-hand. They display and present their evidence as the team of scientists to the 'scientists' conference'.

You could use the following contexts: longevity, size and litter size in animals, diet. Further contexts are suggested in the Key Stage 3 Strategy National Pilot, Unit 4.

You may need to guide the students' choice of enquiry questions so that they are sufficiently focused and linked to the depth of their scientific knowledge. This is an opportunity for differentiation within and between groups.

3.3 Using scientific knowledge and understanding

Students need to apply some scientific knowledge in refining their initial questions. In the examples in teaching suggestion 2 on pages 53–54, knowledge that plants need water, that acid can corrode cloth and that nitrate can be a pollutant underlie the reframing of the question. Students need to apply their developing knowledge of scientific concepts and their developing ideas of evidence throughout their investigative work.

We need to build in focused teaching at various points to teach students:

- what sources of information are available
- how to decide what sources of information are appropriate

- how to search for information in reference books, on CD-ROMs, on databases and on the internet, linking with literacy and ICT
- how to decide whether measurements or qualitative observations are appropriate to a particular piece of work

- how to decide whether it is appropriate to use ICT for collecting data
- how to use preliminary work to find out whether a particular approach is practicable
- how to use preliminary work to help decide what to measure or what to observe, and over what range.

♦ *Teaching suggestions*

1. You can give students practice in identifying key vocabulary in a question (e.g. insulation + heating, lava + shape of volcano, fabric + flame resistance). They can then use an index or web search to find information relating to these words. A homework exercise to collect a paragraph of relevant information for class display can lead to an opportunity to judge which information is relevant to the question formulated.

2. The yeast example given above (see teaching suggestion 3 on page 54) uses scientific knowledge of living things and plant characteristics. You could extend it, as in the original version, to applying ideas of identifying variables. You could use a similar DART strategy to show the students how to use science knowledge in designing an investigation about photosynthesis. Make sure they understand the

target vocabulary by going over the meanings of the words used in the planning activity. Explain to the students some of the harder technical aspects, such as a possible range of methods for collecting gases. Then set them to work on the planning task in Box 3.2.

Box 3.2 *Photosynthesis*

Photosynthesis

You will need yellow and green highlighter pens.
Here is some information:

Green leaves build up energy foods such as glucose when they have light, carbon dioxide gas and water. The process is called photosynthesis. The products of the process are: oxygen gas and glucose. Chlorophyll, which is the green pigment in leaves, absorbs energy from sunlight and stores it as chemical energy. Sunlight is composed of a range of different colours of light. Red and blue light is absorbed during photosynthesis, and green light is reflected. This is why leaves appear green.

What to do

- Highlight information about the inputs to this process in yellow.
- Highlight information about the outputs of this process in green.
- Write a word equation for the photosynthesis process.
- How could the inputs to the process be changed?
- Predict how changing these inputs would affect the process.
- How could the outputs of the process be identified or measured?
- Use this information to formulate a question about photosynthesis that you would like to investigate.
- Would your observations be qualitative or quantitative (measurable)?
- If quantitative, what apparatus could you use to measure the products of photosynthesis?
- Could you use ICT to control the inputs or measure the outputs?
- Exchange this with your partner and evaluate each other's ideas for this enquiry.

3. Whenever students plan enquiries, discuss whether the input variables are categoric or continuous and whether the outcomes are qualitative observations or quantified into measurements. Students can be encouraged to quantify their qualitative judgements, but to beware arbitrary scales which are not continuous. For example four differently shaped forceps can be used to model birds' beaks for picking up seeds. A qualitative judgement of which 'beak' is best can be refined into ranking the four in order, but beware that number 2 is not twice (or half) as 'good' as number 4. Can the students think of a quantitative measure, such as 'how many seeds can each "beak" pick up in a standard time?'

Students make progress from dealing with categoric variables through ordered variables to continuously measured quantities. You can discuss what graphs can be produced from the different kinds of data expressed in words (qualitative) or numbers (quantitative) (Table 3.2). This is discussed in Chapter 4, teaching suggestion 4 on page 79. You can also discuss whether quantifying the data produces better evidence. This can be related to the accuracy and precision of the measuring instruments (see Chapter 2, teaching suggestions 2 on page 26 and 4 on page 27).

You can differentiate the work of students by the extent of quantification and manipulation of the data obtained. You may be able to accept the students' own suggestions or allocate appropriate factors for individuals to investigate. For example, when investigating the factors affecting voltage in a circuit, a teacher asked most pupils to vary the (continuous, quantitative) length of the wire but selected certain pupils to vary the (categoric) material of the wire. When enquiring into the burning of a candle a different teacher expected most pupils to measure the volume of three containers and the time of burning in each. Other pupils 'saved time' by calling their jars 'large, medium and small' but measured the burning time, while the slowest learners just ranked the order of extinction in the three jars. All pupils were satisfied in completing their enquiry and producing data that they could manage to interpret.

Table 3.2 *Expressing different kinds of data*

What we change (*independent variable*)	What we measure (*dependent variable*)	Type of graph
Words	Words	No graph
Numbers	Words	No graph
Words	Numbers	Bar chart

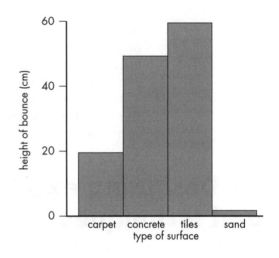

Numbers	Numbers	Line graph

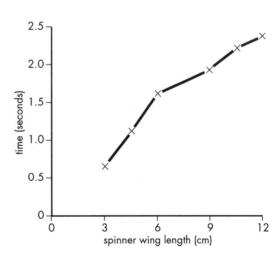

4. Take opportunities to use preliminary work to 'get a feel' of a phenomenon and of the materials of an enquiry, to observe what happens and over what time scale and range of values. When investigating forces, 'playing' with parachutes, balloons, sycamore spinners or trolleys gives an intuitive sense of what to measure.

When studying chemical reactions, a trial with arbitrary (but safely controlled!) quantities of magnesium and acid, or sodium thiosulphate and acid, indicates the range of concentrations and time scale needed. A stroll around a habitat with focused observations can home in on organisms amenable to enquiry. These explorations help to clarify students' scientific knowledge and enable them to plan feasible enquiries. The AKSIS Project *Developing Understanding* Unit 2 'Starting a scientific enquiry: trying things out' gives details of what to do in five different contexts.

3.4 Designing experimental procedures and techniques

Students tend to write endless prose full of unnecessary details about experimental procedure; they associate the investigation process with large quantities of paper and time, and with little thinking. They find it difficult to organise their design. We need to teach explicitly the ideas about evidence involved in the design of investigations. This is discussed thoroughly in Chapter 2. One strategy for identifying variables, defining a question and organising the experimental design is the circle of variables, which is described in Chapter 2, teaching suggestion 2 on page 32. Some alternative strategies are offered in this chapter, both for organising the design of an investigation and as scaffolds for writing a plan in an organised way. Any strategy that you choose should be a tool to teach students:

- how to decide what factors (variables) are relevant to a question
- how to identify independent (input) variables and dependent (output) variables
- whether to make the experiment into a fair test
- how factors (variables) in a situation can be controlled
- how to deal with factors which cannot be controlled.

We also need to teach them:

- about the importance of sample size in biological investigations
- about using controls for comparisons
- how to decide how many observations/measurements to make
- how to choose between different techniques
- what apparatus is available for particular techniques and what to take into account when deciding what apparatus to use
- appropriate techniques for common procedures (e.g. heating solids and liquids, using data loggers)
- how to recognise common hazards in working techniques
- how to recognise and interpret common hazard signs.

◆ *Teaching suggestions*

1. You may like to clarify and consolidate your students' understanding by using the planning poster approach to designing a fair-test investigation. They may be familiar with the system from primary school. It was first published in Goldworthy, A. & Feasey, R. (1994) *Making Sense of Primary Science Investigations*, ASE, Hatfield, and modified in the AKSIS Project *Developing Understanding*, Unit 12 and in KS3 Strategy, Unit 4, Essex posters and computerised versions. You need prepared planning sheets, posters or projection acetates and small self-adhesive strip labels ('Post-its' or similar) (Box 3.3 overleaf).

2. Introduce the question (e.g. 'what is the best way to cure indigestion?') and relate it to the relevant scientific knowledge about neutralisation. Ask your students to identify factors that could affect the efficacy of a stomach remedy and to write each on a sticker (e.g. type of remedy, quantity of remedy, quantity of acid to be neutralised, etc.). They stick these on the 'what we could change' section of the poster, as possibilities for the independent variable (input variable). They can transfer the variable they *choose* to change (e.g. type of remedy). Some may need help in understanding that, although they will not change one remedy *into* another, they will change (substitute) one remedy *for* another in their design. On another colour of sticker students can write all the variables to stick on the section for what they could measure or observe to see whether they make any difference (i.e. possible outputs or dependent variable) and then transfer the chosen one. They transfer all the other variables stickers to the 'we will keep these the same' section.

Box 3.3 *Planning posters (Based on 'Planning posters' Key Stage 3 Strategy Unit 4 National Pilot Scheme (2001) DFES)*

(Level 4–7) <u>**Planning**</u>   **Class group planning board**

Our question is...

We could change   **We could measure/observe**

We will change   **We will measure/observe**

We will keep these the same...

When I change   **What will happen to**

Why?

The two chosen key variables define the question in an operational form. You can reuse the stickers to demonstrate rearranging the variables to plan a different investigation in the context of the original question (e.g. the effect of concentration of a remedy).

3. You could use the computer version of the planning poster, *The Scientific Investigator* originally published by Essex Advisory and Inspection Services and refined in the AKSIS *Developing Understanding* pack. This simulates moving stickers, while using a menu of variables and student's own inputs. The program continues through all the stages of an investigation, allowing students to input and process their data and consider their evidence.

4. Students can progress from physically shifting stickers to writing their variables into a *variables table* (Table 3.3). This serves as a checklist for procedures and a recording chart (see also Chapter 2, section 2.4). Students have to identify the key variables relevant to their investigation as headings to columns on a blank table. Ask them to brainstorm all the 'things' that could vary. They write the variable they want to find out about (outcome or dependent variable) on the last column. All the other variables that might affect the outcome are headings for the other columns. They choose one as the independent variable and then choose values for it. The remainder are control variables.

Table 3.3 *A student's variables table for the investigation: 'what affects the growth of grass seedlings?'* (based on ASE Guide to Secondary Science Education, *p.90, ASE and Stanley Thornes)*

Water	Soil	Size of plants	Amount of light CHANGE (INPUT V)	Growth height (OUTPUT V)
20 ml a day	Litre of soil	2 cm high	Dark cupboard	
20 ml a day	Litre of soil	2 cm high	Natural light	
20 ml a day	Litre of soil	2 cm high	40 W bulb, on 24 hours	

5. You could use a *variables table* or a spreadsheet in a survey (pattern seeking) where you cannot manipulate variables (Table 3.4 overleaf). This helps students to plan which variables to measure and which to attempt to correlate. When they are clear about the design structure of their investigation, students have to plan the practical details. This is where they may need to reconsider their scientific knowledge and do some trials.

Table 3.4 *A variables table for the investigation: 'what factors affect where beech seedlings are found?'*

Seedlings (m^2)	Light intensity	Distance to nearest beech tree	Temperature	Slope	Soil moisture	Soil type

6. When writing up their plans, students could use a range of possible formats. Three of these are:

   - flow chart
   - checklist
   - recipe.

   Students generally understand what is meant by these formats as they are familiar with these structures or scaffolds from other contexts. They are able to use the format to design or report a procedure without a lot of unnecessary repetition or unneeded detail.

   Examples of the above three types of planning format are given in Boxes 3.4–3.6 on pages 65 and 66. You can vary the format that you expect students to use. You could set the task of changing text from one format to another, which students will be familiar with from their literacy experience.

Flow chart

You could write the stages of the plan on cards. Students could then arrange the cards into a flow chart, as in Box 3.4, and then copy it into their books.

♦ As an additional task, students could add to the same flow chart an investigation of different thicknesses and different materials of wire.

Box 3.4 *Flow chart (Resistance investigation)*

How does the resistance vary in different lengths of wire?

Set up circuit:
lab pack, ammeter, voltmeter,
resistance wire

Set the lab pack to 5 V DC

Connect 100 cm of resistance
wire in the circuit

Switch on; record the current
and voltage in the results table

Reduce the length of wire by
10 cm

Is the length of wire more than
25 cm?   Yes

No

Finish practical;
now analyse
results

Checklist

Students could make a checklist of their plan, as in Box 3.5. A tally stroke could be placed in each box as each stage is completed. Alternatively, they could criticise and improve an exemplar checklist.

Box 3.5 *Checklist (Quadrat survey)*

Survey of distribution of plants in school grounds

Assemble the apparatus: quadrat, notebook, pencil. ☐

Select five different grassy areas in the school grounds. ☐

Visit each area; take ten random quadrat readings. ☐ ☐ ☐ ☐ ☐

Record the number of daisies, plantains, dandelions and percentage of grass cover for each quadrat. ☐ ☐ ☐ ☐ ☐

Work out the averages for each area. ☐ ☐ ☐ ☐

Present the results as a wall poster. ☐

Recipe

You could give students a 'recipe' to follow, as in Box 3.6.

Box 3.6 *Recipe*

How do different metals react with acid?

Apparatus
Four large test-tubes
Test-tube rack
Goggles
Bottle of hydrochloric acid (0.4 M)
25 cm³ measuring cylinder
Four pieces of clean fresh metal – iron, magnesium, copper and zinc

Before you start, remember to put your goggles on.
After you finish, remember to wash your hands.

Make sure that your test-tubes are clean and fresh. Place them in the test-tube rack.
Use the measuring cylinder and pour 15 cm³ of the hydrochloric acid into each test-tube (take care – acid is dangerous).
Put one piece of the fresh metal into each test-tube and leave to stand for two minutes.
Record the type of metal in each tube in the results table.
After two minutes, write down careful observations of what you can see happening to each metal.

3.5 Ensuring validity of evidence

Students have to be taught:

- how to make their experiment into a fair test (if relevant)
- how to make all kinds of investigations give valid evidence.

Ensuring that evidence is valid is integral to designing an enquiry and so it is inherent in the items in section 3.4 above. It is featured separately here to emphasise that it is an ongoing process, not finalised with the writing of a plan and setting up the apparatus. When students plan an enquiry to find the answer to a question, they must plan to collect evidence that can provide an answer to that question. This planning for valid evidence is discussed in Chapters 2 and 5. It involves collecting an appropriate quantity and range of data, of sufficient accuracy and reliability. It also involves using appropriate techniques. Although evaluation is often thought to be the 'end' of the investigative process, using evaluation to teach planning helps students to recognise the features of good quality planning. The following activities directly use the skills learned from evaluation to encourage students to grasp the planning skill better.

♦ *Teaching suggestions*

1. You could give the students significantly flawed experimental accounts, or sets of data, such as in Box 3.7. Then ask them to evaluate and improve on the methods used, using some prompt questions. Ask them also to consider the thinking behind the methods.

Box 3.7 *Nasma's work*

Evaporation

I know that liquids will evaporate if left. I wanted to find out about the amount of evaporation from different containers.

I took five different containers. They were:

- saucer
- mug
- milk bottle
- measuring cylinder
- ice cream container.

I put some water in each of them and then I put them where they would be safe.

I put the ice cream container and the milk bottle in the fridge so they would not get in the way. I put the measuring cylinder and saucer on the window sill so they would not get knocked over. I put the mug on my desk because it was safe there.

Later on in the day I looked in each of them to see how much water had evaporated.

> - What question was Nasma's investigation going to answer?
> - What happens when water evaporates?
> - How could you make it evaporate faster?
> - Which of these variables has Nasma kept the same:
>   amount of water
>   temperature of water
>   amount of draughts and moving air
>   surface area of the water?
> - Is this a fair test?
> - How would you change Nasma's method for doing this experiment to make it more fair?
> - How would you measure the amount of water that had evaporated?

2. You could give alternative versions of accounts of experimental plans, for students to compare, criticise and rank. Which methods would give the most useful and valid evidence? (Examples from Chapter 5 could be adapted for this.)

3.6 The whole portfolio of skills at once

It is important to teach that there are several different ideas that go together to make up the ability to plan scientific investigations, so that students realise how to put together a set of separate skills to make a portfolio. Students, both younger and older ones, are too often asked to plan an investigation without knowing what a good plan looks like at their level. An essential step is to show them examples of 'expert planning', but related to their level, approach and vocabulary. There are important teaching points for this. Teach the students how several distinct parts make up the plan. Point out these parts in the plan. Use the language associated with scientific investigations, 'talk the talk' about knowledge, predictions, evidence, fair tests, variables, explorations, validity, reliability, repeats and trial runs. This vocabulary is as much part of the target language of science as words such as photosynthesis and dispersion. Try to keep these 'very expert plans' as user-friendly as possible, so that students will identify with them and copy them. It may help to 'name' them (e.g. 'Nigel's plan'). Explain the plan. Very often teachers give students full and precise instructions about how to carry out a practical task, but omit the step of explaining *why* the instructions are good, so the style can be copied.

◆ *Teaching suggestions*

1. You could use a plan like that in Box 3.8. Get students to underline and colour code the sections that relate to forming a question, to using scientific knowledge, to designing and using techniques and to ensuring validity of evidence. Explain, or ask students to explain, *why* the instructions are good, so the style can be copied.

Box 3.8 *Investigating freezing water and table salt (sodium chloride) (modified from Investigating freezing water and table salt, Hodder Science: Teachers' Book C)*

Salt and freezing water investigation

Question
What does road salt do to ice? Does increasing the salt increase the effect?

Prediction
I know the local council spreads salt and grit on the roads during icy weather. The grit is to give better grip on the road surface. I think the salt dissolves in the dampness on the surface of the road. The salt stops the surface water freezing unless the temperature drops several degrees below 0 °C. Normally, ice forms at 0 °C and makes the road dangerously

slippery. The salt stops this happening. So dissolved salt must lower the freezing point of water. This happens because the salt particles get in between the water particles and weaken the forces between them. This means the water can stay liquid at lower temperatures when the particles are moving more slowly. I predict that the higher the salt concentration, the lower will be the temperature at which the water freezes.

<u>This is what I plan to do</u>
Use this apparatus:

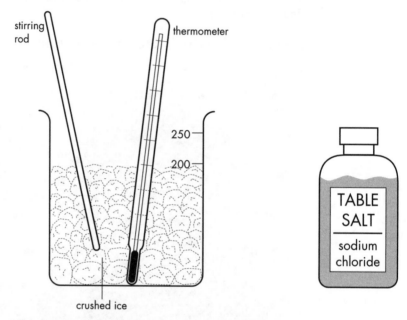

- Fill a 250 cm^3 beaker to the 200 cm^3 mark with crushed ice. I will use this ice for the whole experiment to make it a fair test.
- Measure and record the temperature of the crushed ice to an accuracy of 0.5 °C.
- Add 1 g of salt, stir the mixture, measure and record the lowest temperature reached.
- Record my results on a copy of the table below.
- Add another 1 g of salt (making a total of 2 g). Stir the mixture. Measure and record the lowest temperature reached to an accuracy of 0.5 °C.
- Continue in this way until all the rows of the table are completed.
- Repeat the whole experiment a second time to see if I get a similar pattern of results.

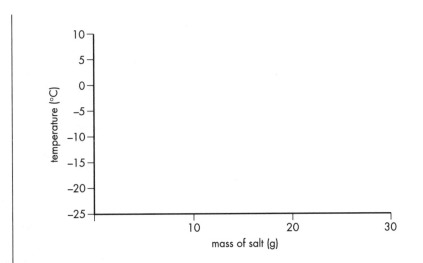

- Then I will plot a *line* graph using the axes above.

Results

Mass of salt added (g)	Total mass of salt in the crushed ice (g)	Temperature of stirred mixture (°C)
0		
I add 1g	1	
Add 1g more	2	
Add 1g more	3	
Add 2g more	5	
Add 3g more	8	
Add 4g more	12	
Add 8g more	20	
Add 10g more	30	

2. You could give a plan with a commentary to introduce target vocabulary and structure the development of planning. Do not flinch from the word count of the commentary. These things need to be explained at length and students are becoming increasingly adept at using text-based resources owing to the success of the literacy strategy. The example in Table 3.6 is incomplete; you need to add your own steps and commentary on filtering, boiling to reduce bulk, evaporating, crystallising and weighing the product to find the yield. (Beware: this investigation is of the 'using and evaluating a technique' kind, so any discussion of variables or fair testing would be irrelevant.)

Table 3.6 *How can we separate rock salt into sand and salt?*

Lennie's version	Added value version
Take about 10g of rock salt and put it in a 250cm³ beaker	We know about dissolving. Salt is soluble and rock is not. So we expect the salt to dissolve and the rock to remain undissolved. Use a top-pan balance. It will weigh accurately to 0.1g. It is not important to get *exactly* 10.0g. But it is important in science to know *exactly* how much you weighed out. It could be 10.3g for example
Add 100cm³ of water to this	The marks on the beaker are only accurate to the nearest 10cm³. A 100cm³ measuring cylinder is accurate to about 2cm³. Smaller measuring cylinders are more accurate
Heat this on a Bunsen burner to about 60°C	Use a thermometer. If you are really careful you can measure temperatures to about 0.5°C. But here you can be approximate
Stir until all the salt has dissolved	

References

AKSIS Project *'Developing Understanding'* and *'Getting to Grips with Graphs'* (details are given in 'Other resources' in Chapter 1, page 16).

Heslop, N. *et al.* (2001) *Hodder Science: Teachers' Resource C.* Hodder Headline, London.

QCA (2000) *Science: A Scheme of Work for Key Stage 3.* Qualifications and Curriculum Authority, London.

Ratcliffe, M. (ed.) *ASE Guide to Secondary Science Education* (1998). ASE, Hatfield, Herts, pp. 89–90.

Science: Tutor's Materials, Key Stage 3 National Pilot, Unit 4 (2001). DfES, London.

The Science Investigator computer program, originally published by Essex Advisory and Inspection Services and refined in the AKSIS 'Developing Understanding' pack. (Details are given in Chapter 1, page 16.)

4 Investigations: Considering evidence

Mike Evans and Linda Ellis

Consideration of evidence is an essential skill in science and is crucial to investigating. Understanding what makes for good quality evidence underpins appropriate measurement, design and recording, so considering evidence has been mentioned in previous chapters. Judging the validity of conclusions drawn will be dealt with in Chapter 5, 'Evaluating evidence'.

4.1 Considering evidence to draw conclusions

Being able to draw a valid conclusion is dependent on proper considering and analysing of evidence and relating it to scientific concepts and hypotheses. Students need to be taught directly how to consider evidence. This is best achieved by focusing first on each of three aspects separately and in isolation from a whole scientific enquiry. Students can then apply what they have learned when engaging in a scientific enquiry.

When considering evidence students need to be taught how to:

- describe relationships by making generalisations about the evidence of patterns in data, including how to spot anomalous results
- draw a concluding remark that directly answers the original enquiry question
- explain their concluding remark by relating it to some scientific model or idea, or suggest further hypotheses as the basis of further enquiry.

◆ Previous knowledge and experience

By the age of 11 students will have been taught the conventions of recording results in tables and of plotting graphs (block graphs, bar charts and line graphs). They will have been expected to interpret these graphs and describe patterns in their results. After investigating the effect of different surfaces on friction, students may well describe the pattern in their results as 'the rougher the surface the more difficult it is to make a shoe slide down a slope'. They may well have been taught to write generalisations as a specific text type, by using comparative adjectives (or words ending in -er). So they may describe their observations in ways such as 'the higher the temperature of the water at the start the longer the water takes to cool'.

They will also have been expected to draw conclusions from their results and begin to relate these to their developing understanding of science. For instance they may write conclusions like: 'Rougher surfaces cause greater friction. This is because friction is caused by bumps on one surface catching on the bumps and ridges of the other. Rougher surfaces have bigger bumps.' When students are studying microorganisms they observe that when yeast is provided with sugar and kept in a warm place it produces bubbles of gas, whereas yeast without sugar produces few bubbles. They may conclude that: 'yeast needs sugar and warmth to grow because yeast is living'. Alternatively when studying changes in materials they may conclude that: 'the reaction between vinegar and bicarbonate is irreversible because a new substance is formed'.

◆ *Teaching to develop these abilities further*

To help students consider evidence we need to build further opportunities to teach them the conventions of how we do this, rather than expecting them to comment on and then assess results. Teaching students about how to consider evidence, as with teaching different ideas of evidence, is best done through short focused activities rather than asking students to perform a whole scientific enquiry first.

Secondary students make progress in their ability to consider evidence by:

- developing their language to describe more complex relationships, including quantitative ones, involving direct and inverse proportionality; using mathematics to describe the relationships and using terms such as exponential and limited growth rate curves
- increasingly drawing on and using scientific knowledge and understanding either to explain conclusions or to ensure that conclusions are consistent with the evidence, for example when faced with sets of competing data

- generating hypotheses from observed patterns
- recognising that the way conclusions drawn from evidence are expressed may differ according to the nature of the enquiry, but that there are conventions associated with these.

4.2 Describing relationships

We need to teach students to see and describe relationships, taking account of discrepancies (anomalous results). We cannot expect students to write generalisations by themselves; we need to show them what we are looking for. You can use the following techniques: reviewing the conventions for describing a pattern or generalisation, modelling generalisations, criticising others' generalisations, improving on others' descriptions, matching generalisations to graphs or to data sets in tables, and discussing anomalous results. Several of these are applications of the strategies described in Chapter 1, page 7.

You can provide many and regular short opportunities for students to review and practise describing patterns. You could do this particularly in advance of sessions where you might want the students to perform a scientific enquiry where they have to describe a pattern in data. Brief activities could be provided as lesson starters or homework activities for later discussion.

◆ *Common problems to address*

When attempting to describe their results some students will merely restate them, making no clear generalisation. For instance when investigating which conditions allowed water to evaporate faster Rob stated:

My results show me that the hottest one evaporated first. The next one to evaporate was the next hottest. The next one was the medium heat one. The last one to evaporate was the coldest. This is what I thought. The reason for this is that hot water evaporates first.

Sometimes when describing more complex relationships by considering more than one variable students ignore one aspect of the evidence. For example, Sara was investigating the factors affecting the size of craters on the moon. She explored the effect of speed. She simulated the effect by dropping balls from different heights onto damp sand. She reported the results shown in Table 4.1.

Table 4.1 *Sara's results*

Results		
Height of drop of ball (m)	Depth of crater (cm)	Width of crater (cm)
1.0	1	10
1.5	2	12
1.75	2.5	13
2.0	3.0	11
2.25	3.5	14
2.5	4	12
3.0	5	14

Conclusion

My results tell me that the higher I drop the ball from, the deeper and wider the crater. This was the same as my prediction. Meteorites travelling at faster speeds would give rise to deeper and wider craters. This was what I predicted.

In this case the student made a generalisation about two variables at the same time, the second variable (width of crater) not being consistent with the evidence, thus leading to a wrong conclusion.

◆ *Teaching suggestions*

1. You could teach directly through a starter activity, by providing a series of different data sets, as in Tables 4.2 and 4.3.

Table 4.2 *Animal gestation periods*

Animal	Gestation period (months)
Elephant	22
Horse	11
Human	9
Baboon	6
Otter	2

Ask your students to describe the pattern they see. For Table 4.2 they should be encouraged to say 'The bigg*er* the animal the long*er* the gestation period'.

For Table 4.3 a description of the pattern would be 'The faster the speed the greater the braking distance'.

Table 4.3 *Braking distances*

Speed of car (mph)	Braking distance (m)
20	13
30	21
40	40
50	60
60	80
70	105

For a 'context-free' and interactive exercise to assimilate this form of statement, you could use 'Linking the actions' from the computer disc of *Getting to grips with graphs*. Several books provide a good range of data and exercises that allow students to practise describing relationships, for example *Getting to Grips with Graphs*, *Thinking Science* (CASE) and *Data Handling Skills for GCSE* (Cambridge).

2. Once students have understood the point about using comparative adjectives you might increase the difficulty by providing them with result tables (e.g. Table 4.4) that first require ordering, thus making the point that results tables need careful construction to help us see the relevant patterns.

Table 4.4 *Weight and gestation periods*

Animal	Average weight of offspring at birth	Average gestation period
Guinea pig	100 g	68 days
Dog	230 g	2 months
Mole	1 g	30 days
Lion	1.3 kg	106 days
Elephant	110 kg	22 months
Giraffe	60 kg	450 days
Mouse	1 g	20 days

Ask these questions: What pattern do you see? Can you describe it? Do you need to do something with the data first? What do we need to remember when creating sets of data?

3. You could apply the same principles when asking students to describe patterns in graphs. Looking at a range of line graphs and using overhead projector (OHT) overlays can provide good opportunities for students to practise describing patterns in graphs and to become familiar with the shape of graphs (Figure 4.1).

OHT with blank axes:

Selection of overlays with different labels, e.g:

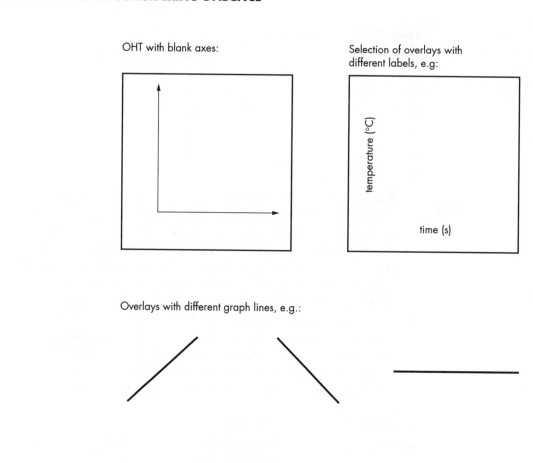

Overlays with different graph lines, e.g.:

Figure 4.1
Graph OHTs and overlays.

You can practise 'as the . . . so the . . .' (using comparative adjectives), or for example 'as the time increases so the temperature decreases' or, with the slope facing the other way, 'as the time increases so the temperature increases'.

AKSIS *Getting to Grips with Graphs* provides further opportunities for students to develop their descriptions of patterns. Unit 11 'Sketch the line' provides some good examples of how this can be developed by asking students to think about relationships and to sketch the line they would expect. They are encouraged to sketch the line they think should represent the relationship between the variables in Figure 4.2.

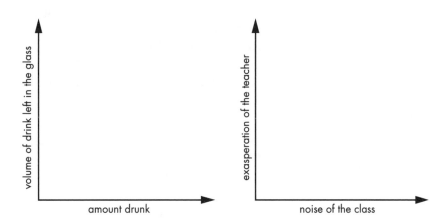

Figure 4.2
Drinks and noise.

Younger students should be taught ways of describing the shapes of graphs so that they develop a sense of what the data are representing. *Getting to Grips with Graphs* Unit 9 'Describe the pattern' encourages them to do this using everyday language. As students progress you will want to teach them to recognise and describe relationships in line graphs appropriately as directly proportional or inversely proportional.

4. Students need help in deciding what graph to plot (i.e. is it a bar graph, a bar chart or a line graph?). Richard Gott and Sandra Duggan point this out in their book *Investigative Work in the Science Curriculum* (OU Press). They emphasise that students need to understand that data (or variable) type dictates directly the type of graph plotted and that students need to be taught the connection. Table 4.5 should help.

Table 4.5 *The three basic graph types result from plotting data against data*

Data type	Data type	Resulting graph
Discrete or categoric	Discrete or categoric	Block
Discrete or categoric	Continuous	Bar chart
Continuous	Continuous	Line graph

You could give students practice in describing the relationships illustrated by the different kinds of graph. Help them with ways of describing the types and shapes of graphs so that they develop a sense of what the data are representing.

The type of graph to be planned and the evaluation of the choice of graph are discussed in teaching suggestions 3 on page 58, and 1 and 2 on page 99 respectively.

5. As students progress in their understanding, so they will need to be taught to describe increasingly complex relationships and to recognise curves and mathematical relationships. You could provide students with sets of data for them to attempt descriptions of data that are not directly proportional, like those in Table 4.6.

Table 4.6 *Sunshine data*

Month	Amount of sunshine (hours)	Month	Amount of sunshine (hours)
January	1.5	July	6.6
February	2.4	August	5.6
March	3.7	September	4.5
April	5.0	October	3.2
May	6.2	November	1.7
June	6.7	December	1.4

Help students to make statements such as 'In this particular year the amount of sunshine increases from January to June and then decreases again'.

The use of data sets like this can also be used to encourage them to think carefully about how they might better represent the data as graphs so that patterns can be spotted and described more clearly.

6. Modelling or criticising others (see Chapter 1, page 7) is a very good way of showing students what is expected and helping them to understand what is meant by 'the quality of expressing relationships'. You could collect together some examples of generalisations (descriptions of relationships both good and bad), perhaps from previous students, and read through some together. You could identify the good and bad features. You could also draw out the principles of using comparative adjectives.

For example, you could discuss with students the merit of each of the following examples:

(i) My results tell me that as you change the number of turns in an electromagnet from 10 to 20 and 30 and so on the number of paperclips it picks up is first 3 then 5 then 8 then 12 and finally 26.

(ii) My results tell me that, in general, the greater the number of turns the greater is the magnetic effect.

Students could also, in groups, discuss other examples, perhaps trying to list them in rank order from the best to the worst and trying to give their reasons. This method is used in Unit 14 'Describing relationships or patterns' of AKSIS *Investigations – Developing Understanding.* For example, in activity 14d students are invited to look at the recorded data resulting from investigating the connection between the mass of wood mice and the length of their tails (Table 4.7). You could generate your own data by exploring the connection between finger length and height.

Table 4.7 *Mouse tails (from* Developing Understanding, *p. 83)*

What pupils said	Does it mention both factors? (YES or NO)	Does it describe the general pattern? (YES or NO)	Does it indicate that data from some individuals do not fit the pattern? (YES or NO)
The one with the shortest tail was worst			
All the mice with short tails have little mass			
As the wood mice tails get longer their mass increases			
In general, the longer the tails the greater the mass			

Figure 4.3
Finger length and height.

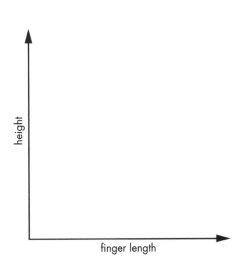

7. You will need a range of opportunities to develop students' ability to consider increasingly complex relationships and consider what constitutes a pattern, including what constitutes an anomalous result. You could ask students to improve descriptions written by others, matching descriptions with patterns and considering anomalous results. Unit 14 of AKSIS *Developing Understanding* provides some examples, using scattergrams in which anomalous results can be considered (e.g. 14d as mentioned above). Scattergrams can be generated for a number of relationships that are likely to provide anomalous data, such as investigating the relationship between finger length and height for the students in the class (Figure 4.3).

4.3 Drawing concluding remarks

After having helped students to describe a pattern we need to encourage them to make a concluding remark that relates back to the enquiry question. This often involves moving them from describing relationships, but in some cases conclusions may be drawn directly from observations and measurements. In any case the conclusion must be consistent with the evidence. This step is not too difficult to teach, but is often missed by students, so they do not tell the whole story and the writing can look incomplete. You can use the following techniques: model conclusions discussing what makes good conclusions, review and criticise conclusions written by others, discuss whether conclusions drawn by others are consistent with evidence, or discuss the conclusions drawn at different times in history on the same evidence.

◆ *Common problems to address*

After describing a pattern in results many students do not then go on to make a concluding remark that relates their observations back to the original line of enquiry. For instance, when a group were investigating 'the factors affecting the heat obtained from a range of fuels', they explored the viscosity, density and number of carbon atoms present in a molecule of different alcohols. They described the patterns 'the more dense the fuel the more heat is produced, the more carbon atoms the more heat is produced, the more viscous the fuel the more heat is produced'. They did not attempt to draw these ideas together by attempting to relate it back to the question. They could have said 'I conclude the main factor affecting the heat produced is the size of the molecule' and then go on to explain why they could link the three aspects together.

◆ *Teaching suggestions*

1. This is a place for direct teaching. You could draw students' attention to examples of questions, or their own enquiry question, and model the concluding remark. An example is given in Table 4.8.

Table 4.8 *Concluding remarks*

Enquiry question	Patterns described	Concluding remark
What affects the solubility of salts?	Nitrates are soluble Sodium and potassium salts are soluble	The type of acid the salt is made from and the metal affect the solubility
	Carbonates are insoluble	Generally all nitrates and sodium and potassium salts are soluble and carbonates are not
What affects the strength of an electromagnet?	The greater the number of turns of wire the stronger is the magnet The larger the current the stronger is the magnet	The strength of the electromagnet depends on the number of turns and the size of the current Generally the higher the current . . .
What affects the pressure of a gas?	The hotter the temperature the greater is the pressure The smaller the volume the greater is the pressure	The pressure exerted by a gas depends on its temperature and the volume Generally . . .

Subsequently, you could model the concluding remarks to another enquiry question by taking suggestions from the class and building up a consensus statement on the board. Discuss what is good or bad about the accepted or rejected contributions from individuals.

2. You could provide short activities allowing practice for writing conclusions from patterns in data. You could make cards with patterns in data shown on cards of one colour and conclusion statements on cards of another colour. The task for students is to match the conclusion to the correct pattern.

3. You could expose students to the ways in which conclusions are written by others, for example other students, in newspapers, scientific journals and publicity materials for pressure groups. Discuss whether the conclusions are consistent with evidence, using topical scientific debates (e.g. BSE, whether badgers spread tuberculosis, or the disappearance of song birds). This approach is extended to evaluation in sections 5.8, page 105 and 5.10, page 106. (See also Chapter 9.)

4. Resources from the history of science can be used. You could discuss the conclusions drawn at different times in history on the same evidence – for example the advice given about diet and weight loss by the Consumer Association in 1957 compared with now, or conclusions about magnetism. (See also Chapter 8.)

5. Whenever you make direct observations and measurements in any practical work, you can help students to practise making direct conclusions. Ask them to think whether questions can be answered purely on the direct evidence, for example 'what happens when we heat iron filings and sulphur?', 'which is the most reactive of these metals?', or 'which are the commonest weeds in the flowerbed?' Practise writing conclusions based on direct observations and measurements from these enquiries.

4.4 Relating conclusions to scientific knowledge and understanding

To be effective and to motivate students we would normally expect them to draw on their knowledge and understanding to make a prediction about what they would expect to happen. So attempting to explain a remark using scientific models or theories should not be an afterthought. Rather it should, wherever possible, be the starting point and may appear in plans. However, you will need to be careful since students may not have the knowledge they need to explain an idea. Where appropriate, students can be asked to add to the concluding remark 'This is because...'. At Key Stage 3 we should concentrate on helping them to describe relationships and to use scientific understanding of the key ideas of particle theory, energy transfer, force and cell to explain conclusions.

Leading students, either to explain the remark by relating it to some scientific model or idea or to suggest hypotheses that would form the basis of further enquiry, you could use the following techniques:

- considering evidence needed to support competing theories
- considering the conclusions drawn by different interest groups
- judging the quality of others' explanations for an observed pattern
- practising relating conclusions to scientific knowledge through pairing activities.

♦ *Common problems to address*

In the example using alcohols on page 82, the students had been taught the structure of alcohols and so they would have the scientific ideas and models to help them explain why the size of the molecule was the factor affecting the production of heat. However, when pupils attempt to use their scientific knowledge to explain patterns, too often the knowledge is beyond them. This has a significant implication for teaching. If you want students to explain their conclusions using scientific ideas and models then you need to select the enquiry question or the data with care.

For instance, younger secondary students could investigate the factors that affect the swing of a pendulum but would not have the scientific knowledge or ideas to explain their conclusion. Older students could investigate how copper sulphate affects the rate of chemiluminescence in luminol but could not readily explain their conclusion.

♦ *Teaching suggestions*

1. You could use concept cartoons to consider the evidence needed to support competing theories, as explained in Chapter 1, page 10.
2. Published material from interest groups (e.g. Greenpeace) provides sources of conclusions related to scientific knowledge. You could consider the different conclusions drawn by different interest groups (e.g. fishermen and Greenpeace) and how they have related their claims to scientific knowledge.
3. You can help your students to practise judging the quality of others' scientific explanation for an observed pattern by the strategy of criticising other people's work. AKSIS *Developing Understanding* Unit 15 provides examples in several contexts.
4. Your students could practise relating conclusions to scientific knowledge through activities such as 'statements into explanations' and 'true/false pairs', which could be prepared as card sort exercises.
5. You could consider examples where students draw conclusions directly from observations or evidence by relating them directly to existing scientific knowledge. For example, they could try deducing the identity of a chemical substance from a description of its properties, which depends on knowledge and understanding of the properties of groups of chemicals.

4.5 Putting it all together: writing conclusions

In drawing conclusions from their whole scientific enquiries, we want students to consider results, look for any patterns, describe the pattern as a generalisation using comparative adjectives, say whether it was expected or not and try to attempt an explanation of the generalisation from what they have learnt previously.

The key questions to ask are:

- Can we answer the question directly from observations and measurements, or do we need to look for a pattern?
- Is there a pattern or not? What do the results tell you?
- How can you describe the pattern (as . . . so . . .)?
- Is this pattern expected?
- How do the data (or pattern of data) answer the question of the enquiry?
- How can you explain the pattern?

◆ *Teaching suggestions*

1. After having collected results, you could guide students through the sequence by inviting them to spot patterns and then collectively, on the board, write the conclusion. Get students to read this out loud before rubbing it off. Then ask them, either in pairs or individually, to write their own, bearing in mind what was discussed.
2. Model a good conclusion by talking through the process with students.
3. Following the modelling of a good written conclusion to an investigation, you could collect together some examples of conclusions (good and bad) to this investigation. Read through some of these together. Point out the good and bad features and draw out the common elements of good conclusions. Then give the students some 'good' results and ask them to write a conclusion to the results.
4. You could set students a task to write a conclusion for a set of results then ask them to pair up and compare their conclusions. Are they the same? Do they agree? Which parts do they agree with? Which do they not? Can they come up with an improved version? Then provide them with a checklist as in Box 4.1 to help compare and assess their original and final attempts.

Box 4.1 *Checklist*

> - Is there a pattern in your results?
> - What is the pattern?
> - Is this what you expected?
> - Can you explain the pattern?

◆ *Using supports such as writing frames*

We need to help students realise that in the main a 'conclusion', the outcome of considering evidence, is a text type. Text types are taught to students as part of their studies in literacy. By convention this text type includes:

- a description of the pattern
- a concluding remark (summarising the findings)
- a reason for the conclusion.

You can directly teach the conventions of writing conclusions using writing frames to support students' early attempts. The use of such supports is discussed in Chapter 1, section 1.3.

4.6 Summary

With younger students we should concentrate on helping them to describe relationships and to use scientific understanding of the key ideas of particle theory, energy transfer, force and cell, to explain conclusions. We should also develop their understanding of the importance of evidence and the conventions of writing conclusions. To enable progression for older students we should focus more on developing their ability to look for evidence to justify conclusions, describe more complex relationships, and relate a wider range of scientific ideas to their conclusions. We need to help students realise that in the main a 'conclusion', the outcome of considering evidence, is a text type. Throughout, we should reinforce the conventions of this text type when writing conclusions.

However, the purpose of considering evidence should not be reduced to a formula; students need to consider their evidence in relation to prior and subsequent planning and evaluation.

References

AKSIS *Developing Understanding, Getting to Grips with Graphs* (see 'Other resources' in Chapter 1, page 16 for details).

CASE *Thinking Science* (see Chapter 1, page 16).

Gott, R. & Duggan, S. (1995) *Investigative Work in the Science Curriculum*. OU Press, Buckingham.

Pritchard, I. (1989) *Data Handling Skills for GCSE*. Cambridge University Press, Cambridge.

Scientific Investigations packs. Collins, London (see Chapter 1, page 16).

5 | *Investigations: Evaluating evidence*

Rod Watson and Valerie Wood-Robinson

To think as scientists and informed citizens, students have to make judgements about the reliability of evidence and validity and strength of conclusions: they have to be able to evaluate. This depends on understanding the ideas of evidence discussed in Chapter 2 and using them to evaluate scientific enquiries.

5.1 Evaluation, the key to enquiry

Evaluation is not the end of an enquiry. It is part of an iterative process, leading to further planning, refining the strategies for collecting evidence and making further measurements or observations, and it may take place as the enquiry is proceeding. Evaluation assesses how well the purpose of the enquiry has been fulfilled. It is concerned with how trustworthy the evidence is and whether it supports the conclusions drawn. There is no one template for evaluating: the criteria for judgements vary with the kind of enquiry.

When evaluating evidence, students need to be taught to:

- examine data critically to consider its reliability, identify anomalous data and look for evidence of bias
- examine the processing and representation of the data
- judge whether the evidence is sufficient to support the conclusions, and whether these make use of all the evidence
- judge whether the experimental design was appropriate and the method was suitable to make the evidence valid
- having considered these aspects, make suggestions for improvements to the methods and further investigations.

◆ *Previous knowledge and experience*

'Evaluation' is used for a number of different judgement processes both in education and in everyday life. In the context of scientific enquiry, however, it should be used to judge the quality of evidence and the conclusions drawn from that evidence. By the age of 11, students will have some experience in using 'evaluation' in this way. Most will have used evaluation to decide whether their conclusions agree with any prediction made and whether they enable further predictions to be made. They will have reviewed their work and the work of others and described its significance and limitations. When carrying out fair tests they will have considered whether they have controlled relevant variables.

Some students, however, may evaluate their performance or enjoyment of their work, or may uncritically congratulate themselves on confirming their prediction. They will understand everyday meanings of words like fair, reliable, error and uncertainty, but will not easily distinguish the technical sense of their use in evaluation. When suggesting how to improve their enquiries, many students think that 'more' must be better. They think they should take more readings, more samples, consider more variables, be more careful, and more organised, and use more sophisticated equipment, rather than identify particular readings, ranges or strategies that need to be improved.

Even within science, students may have used 'evaluation' in different ways, for example to evaluate a product or process in a technological enquiry (e.g. of the effectiveness of stomach powder or of methods of drying an apple), or to criticise a piece of writing or a poster. These judgements of quality differ from the evaluation of scientific evidence, such as judging whether sufficient data have been used as a basis for developing an apple-drying process in an investigation of the water content of apples. In this chapter, activities are described that help students to be much more specific in their evaluations of scientific enquiries and to relate their evaluations to the quality of evidence and its interpretation.

Some students are biased when evaluating evidence. They want to be *right* to satisfy themselves, outdo their friends and please their teachers. They look for evidence to confirm their predictions. They tend to ignore, dismiss or fail to observe contradictory evidence. To place them at a distance from the evidence collected many of the teaching suggestions that follow involve students evaluating other people's work.

♦ *Teaching and assessing evaluation*

Evaluation can be taught and assessed during learning:

- through teaching activities focused on specific aspects of evaluation (see sections 5.2–5.5)
- through teaching within a class or individual enquiry (see sections 5.6–5.9)
- by teaching critical judgement of secondary sources of evidence (see section 5.10).

The criteria for judgement of reliability and validity vary with kind of enquiry, and so there is no one prescribed formula for evaluation. For example, questions about controlling variables are relevant in some kinds of enquiry (fair testing), whereas questions about sample size are important in others (pattern seeking). When discussing evidence with the class, the key question is whether the evidence can be trusted. Students need to be taught to examine raw data critically and to question how it was obtained. Questions that can be discussed include the following:

- How sensitive and accurate were the instruments, and how accurately were they read?
- Was that degree of accuracy appropriate to the task? What was the uncertainty or error in the readings?
- What caused anomalous results and should anomalous results be included in the data or rejected?
- Have sufficient data been collected to judge their reliability?
- Have the data been recorded to an appropriate number of significant figures?
- Have the data been processed appropriately to cope with error (e.g. by averaging or by using graphs)?

These questions are addressed in planning an enquiry (see Chapters 2 and 3) and in collecting and analysing data (see Chapter 4). They need to be revisited to evaluate the procedure and conclusion of the enquiry.

5.2 Accuracy and reliability of data

If a person is reliable you can trust him to be consistent and to tell you the truth. This is a good starting point for an analogy with reliability of evidence. Reliable evidence is consistent and gives a 'true' description of the matter under consideration. Whether the evidence can be trusted depends on the sufficiency of the range, interval, and number of readings of different values, on the number of repeat readings of the same values and on how the repeat values are processed.

If repeat readings are closely clustered, then the results are pretty reliable, just as the existence of many witnesses giving similar evidence increases its reliability. The more repeat readings we take, the more information we have about reliability. However, it is possible to get results that are repeatable but which are inaccurate, for example a student may get consistent readings from a wrongly calibrated thermometer, so it is also useful to compare results obtained by different people doing the same experiment.

Older students may progress to a deeper understanding of the meaning of the concept of 'uncertainty'. The single value obtained at the end of a measurement process is only an approximation of its true value. Uncertainty is a statistical parameter, based on the data collected, that is used to describe the limits within which the true value is expected to lie. The term 'uncertainty' is preferred to the term 'error' by some teachers, as students use the word 'error' to refer to mistakes, accidents or miscalculations, not to the concept of experimental error leading to a range of uncertainty.

Consideration of anomalous results and trying to explain them is important in evaluating the credibility of evidence and in promoting further hypotheses and enquiries. An anomalous observation or measurement may not be obvious; judging it as anomalous depends on recognising the range of uncertainty.

◆ *Teaching suggestions*

1. You could use the following example of student work to consider the reliability of results and their interpretation.

    Students had been investigating the time taken for different kinds of paper to burn completely. They predicted that the time taken depended on the weight of the different kinds of paper. They had used four kinds of paper, cut into the same shape and size. They ignited them and held them in the same way whilst the paper burnt. Box 5.1 shows Sam's work.

Box 5.1 *Sam's work*

Results

Kind of paper	Weight (g)	Time for trial 1 (s)	Time for trial 2 (s)	Time for trial 3 (s)	Average time (s)
Newspaper	2.03	20.23	18.13	22.33	20.23
Ordinary white paper	3.05	33.62	30.44	30.31	31.45
Wrapping paper	4.10	48.88	47.07	44.84	46.93
Graph paper	4.52	46.31	44.03	48.02	46.12

Interpretation

The results show that the time taken for the paper to burn depended on its weight. The wrapping paper was an exception. It took longer to burn than we expected.

Students need to ask such questions as:

- Did the repeat readings give very different results?
- Has Sam taken enough repeat readings to tell whether his results are reliable?

One way of using these data to examine the idea of accuracy is in a whole class demonstration and discussion. You could ask different students to weigh the same piece of paper and record the result without seeing one another's results. Next, burn the paper in front of the class whilst several students with different stop-watches time the burning. Then compare the results that have been collected and use these to consider the reliability of the data and whether the data in the table above should be recorded to such a degree of accuracy. This could lead on to considering whether you can really say that there is a difference in the burning time of the wrapping paper and the graph paper. Finally you could consider how more trustworthy data might be collected.

2. You could ask students to consider evaluations written by other students (see Chapter 1, page 7). You could also use the students' evaluations of the burning paper enquiry in Box 5.2.

Box 5.2 *Students' evaluations of the burning paper enquiry*

Daniel's evaluation

We did three trials for each kind of paper. Because we have lots of readings and took an average our results are reliable.

Laura's evaluation

We did two trials for each kind of paper. There was never much difference between the results for each kind of paper so we think the results are reliable.

Kirsty's evaluation

We did three trials for each kind of paper. The spread of the results was about 4 seconds, which is about a 10% error. This means that we cannot detect a difference in the burn times of the wrapping paper and graph paper. We would have to do a lot more trials and take the average to see whether there was a difference.

Students could work in groups and put these statements in order of the quality of evaluation. Once they have done that, they must give reasons for this order (i.e. they must make explicit the criteria that can be used to judge the quality of an evaluation). Daniel's evaluation simply describes an acceptable procedure for dealing with error without discussing the spread of readings. Laura's evaluation provides some justification for thinking the results are reliable, whereas Kirsty's evaluation includes an estimate of uncertainty. You can use the same strategy, with more sophisticated statements, to enable older students to progress to a deeper understanding of uncertainty.

3. You can also select student evaluations to focus on the procedures used in carrying out the enquiry, as well as the error inherent in making measurements. The examples in Box 5.3 are from an investigation of the effect of temperature on the rate of reaction of sodium thiosulphate solution with dilute hydrochloric acid. Students mix the two solutions together in a conical flask at different temperatures. They place the conical flask on a cross drawn on a piece of paper and time how long it takes for the cross to disappear from sight when viewed through the solution.

   You could show students how this experiment is carried out and then ask which of the evaluations in Box 5.3 is the best, and why. By doing this students make a decision about the quality of the evaluation and then make clear the criteria that they have used to judge that quality.

Box 5.3 *Students' evaluations of the effect of temperature enquiry*

Ben's evaluation

There was error in measuring the temperature and measuring the time. The thermometer measured the temperature to one degree and when we started timing the temperature was quite accurate, but with the hotter mixtures the flask cooled as we were doing it. This was the main source of error. The stop-watch timed to one hundredth of a second so that was really accurate.

Nick's evaluation

Although the stop-watch was really accurate, it was difficult to decide when the cross had disappeared, so the timing could be out by as much as 2 seconds. We found that the temperature at the beginning and end of the experiment was different so we decided to take the temperature at the beginning and at the end and find the average. A better way of doing the experiment would have been in water baths set at different temperatures. One result in our graph is way out and we think this is because we forgot to wash out the conical flask between experiments.

Julie's evaluation

We predicted that the reaction would get faster as it got hotter and we were right, so we did well. We did the reaction three times at each temperature and worked out the average time using a calculator. It gave an answer to four decimal places so this is very accurate.

Claire's evaluation

When we measured out the quantities of solutions into the conical flask, we measured the volumes using a measuring cylinder, so that was accurate. It would have been better if we had used the same person to time the experiment each time.

Using statements such as these, students can be guided to consider the various different sources of error, the significance of each and how these errors could be combated by using different measuring instruments, using different procedures or by repeating readings and averaging. You could prepare different sets of 'student' statements to focus on different aspects of evaluation. For example they could be used to focus attention on the selection of suitable measuring instruments for volumes and making an estimate of the amount of error (Box 5.4).

Box 5.4 *Students' evaluations of volume measurements*

Selma's evaluation

All the volumes were measured carefully. We used the same volumes of sodium thiosulphate solution and acid each time and we measured them carefully using the 50 mark on the beaker.

Pippa's evaluation

We measured the volumes of the solutions using a $250\,cm^3$ measuring cylinder. This measured to an accuracy of $5\,cm^3$ so our readings might have been out by about 2 or $3\,cm^3$. It would have been better to use a measuring cylinder that measured to an accuracy of $1\,cm^3$.

Joey's evaluation

We made the measurement of volumes accurate by using a measuring cylinder that had marks for every $1\,cm^3$. This meant that when measuring $100\,cm^3$ we had an error of less than 1%. We think that this was accurate enough for this experiment.

These statements can be used to consider the accuracy of the measuring instruments, the significance of errors in the reading of the volume on any patterns to be observed in the enquiry and, for more advanced students, quantitative estimates of error.

4. Identification and explanation of anomalous results are important both for evaluating the credibility of evidence and for promoting further hypotheses and enquiries, but an anomalous observation or measurement may not be immediately obvious.

   You could use the data set in Box 5.5 to teach how to identify anomalous data. It was recorded by a group investigating parachutes. The students knew from the outset that if you drop parachutes from different heights they take different times to fall; the bigger the drop the longer would be the time to fall. They were uncertain about the pattern of these results, however. Some predicted a graph that would be a straight line going through the origin whereas others thought that it would be a curve. Looked at critically, the data set shows anomalous results that do not fit with patterns and 'spoil' averages. The task asks the students to use two stages in identifying anomalous data: first scanning the sets of three values in the table and second plotting a graph using all the data, not just the average time.

Box 5.5 *Student's exercise book: parachutes*

Study this extract from a student's exercise book and then answer the questions below:

I dropped my parachute from different heights. I tried to get it to work properly but sometimes the string tangled up.

Results

Height of drop (m)	Time for first try (s)	Time for second try (s)	Time for third try (s)	Average time (s)
1.0	1.04	1.50	0.99	
1.5	1.61	1.64	1.59	
2.0	2.25	2.22	2.17	
2.5	2.31	2.48	2.84	
3.0	3.41	3.38	3.50	
4.0	3.64	4.61	4.54	

Questions

1. Look at the table. Which results seem to be very different from the others for each height? Circle in blue each result that you think might be very inaccurate.
2. Plot the results as a graph using all the values in the table. Do not work out the average yet, just use the readings for each try and plot them all on the same graph. Draw a line of best fit.
3. Look carefully at the graph and circle in red three points that do not fit the pattern.
4. Now look at the table and circle in red the same three pieces of data.

Here is an example of the graph drawn by students:

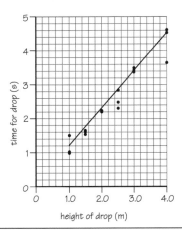

The plot indicates a straight line and confirms that the value of 3.64 s for 4.0 m does not fit. The three points plotted at 2.5 m show that one value is close to the line of best fit (i.e. 2.84 s) and that this time two values do not fit the pattern (i.e. 2.31 s and 2.48 s).

(For a further example see AKSIS *Getting to Grips with Graphs* Activity 20: 'Plotted points exact or approximate'.)

You can now focus the class discussion on how to deal with anomalous data. Going back to the table, there are two ways of coping with the anomalous data:

(i) Choose the middle value of the three values to plot as a simple way of eliminating most anomalous data. The middle value is more likely to be closer to the 'true' value.

(ii) Anomalous data can be identified and then excluded from the average. In the more unusual case where two out of three values are anomalous, we are left with only one value that can be plotted – pointing to the need for more repeat readings.

5. An alternative way of dealing with anomalous data is detailed in Activity 24 from *Getting to Grips with Graphs*. In this activity, students are supplied with graphs in which there are anomalous points. Drawing a line of best fit involves deciding whether a straight line or a curve is more appropriate. The best way to do this is by relating the graph to the real situation which it is describing. For example, should the points on a temperature–time graph of a liquid cooling be a straight line or a curve? Relating the graph to the real physical situation helps to decide this, and hence which point on the graph may be anomalous; it simply does not make sense for the cooling curve to go below room temperature, so the graph must reflect this.

5.3 Representing data to allow us to detect patterns in evidence

Another aspect to evaluate is whether the appropriate data have been selected for display, for example whether to plot all repeated points or the mean or the 'best', and whether an appropriate kind of graph has been selected (e.g. scattergram, bar chart or line graph). These points are discussed in other chapters (e.g. teaching suggestions 5 on page 27 and 4 on page 79).

♦ *Teaching suggestions*

 1. You could give your students examples of other students' work for them to evaluate for appropriate representation of evidence, including choosing the right kind of graph, scales, line of best fit, etc. *Getting to Grips with Graphs* includes activities on 'Spot the mistakes' and 'Does it make sense?' both in the printed materials and the IT discs; *Science Investigations* also contains useful examples.

2. Students should examine their own graphs to see whether they 'make sense'. They could be asked the following questions. If the graph goes through the origin (0, 0) what does that point mean? Could it be a real value? If you read off intermediate (interpolated) values from your line graph, do they make sense, or are you trying to read a value between two discrete values such as 'carpet' and 'tiles', or 'red' and 'blue'? If they don't make sense, have you drawn the wrong kind of graph?

5.4 Judging the sufficiency of evidence and full use of the evidence

The amount of data required to provide reliable evidence varies with the enquiry. In a fair-test enquiry, the number of values of the variables and the range and interval need to be examined to judge whether they provide sufficient points to indicate a pattern. In some cases three or four points over a narrow range might be sufficient, whereas in others a large number of points at narrow intervals over a wide range would be needed. It depends on the reliability of the results and the complexity of the pattern being shown in the graph. In a correlational survey or pattern-seeking enquiry, evaluation of sufficiency means judging whether the sample was sufficiently large and random.

♦ *Teaching suggestions*

Activities that consider range, interval and number of values, in the context of *planning* a fair test investigation, could be adapted to represent the outcomes of enquiries and thus serve as material for *evaluation*. Items in Chapters 2 and 3 could be revisited.

1. You could ask your students to look at the results of an investigation into shadows shown in Box 5.6.

Box 5.6 *Investigation into shadows*

Paul's table

Distance between light and pencil (cm)	Length of shadow (cm)
10	47
20	29
30	24
40	20
50	19

Rita's table

Distance between light and pencil (cm)	Length of shadow (cm)
50	19
100	16
150	13
200	10
250	8

Mal's table

Distance between light and pencil (cm)	Length of shadow (cm)
10	50
33	28
40	25
68	18
70	17

Which set of results gives the most useful evidence about how the length of a shadow varies with distance between the object and the light? Evaluate each student's choice of values of independent variable, and say whether sufficient evidence was obtained and whether all the values provided useful evidence.

2. You could use examples of students' statements that evaluate the sufficiency of evidence, as in Box 5.7.

Box 5.7 *Students' evaluations of evidence sufficiency*

> Students thought that daisies survived better on the main field than on the football pitch. Did they have enough evidence to support this?
>
> Lucy   Yes. I saw more daisies along the edge of the main field than on the football pitch.
>
> Dennis   Yes. We counted more daisies in a quadrat one metre square on the main field than in one metre square on the football pitch.
>
> Rene   No. We didn't have time to count every daisy plant on both fields.
>
> Ravi   Yes. We counted the daisies in ten separate metre squares scattered over each field, and found more daisies on the main field.
>
> Put these answers in order from best to worst. Explain how you decided the order.

5.5 Judging the validity of the evidence

A fair-test investigation is one in which we change something (the *independent variable*) and observe the effect that it has on something else (the *dependent variable*), while controlling all other relevant variables at constant values. Evaluation of 'fairness' is important in establishing *validity*, that is how sure we can be that the effect was due to the factor we were investigating. It involves asking these questions: how easy or difficult was it to identify factors to be controlled? Which factors were not controlled and made the investigation 'unfair'? How would this affect the results? How could you improve the 'fairness'? What could you do about factors that you could not control? It also involves weighing up the demands of stringent fair testing against feasibility.

◆ *Teaching suggestion*

You could ask students to comment on other students' evaluations of 'fairness' as in Box 5.8, which evaluates the effect of temperature on the time taken to dissolve sugar.

Box 5.8 *Students' evaluations of 'fairness'*

Melissa

We each had a go at each job, nobody in our group was left out so we were very fair to everyone.

Olivia

We did each temperature three times and took the average. This made it a fair test.

Aidan

We did a fair test because we kept all these the same: the volume of water, the number of stirs, the person doing the stirring and the amount of sugar.

Finley

We tried to keep the volume the same, but it is difficult to get the water exactly to the level line. We tried to weigh out exactly 10 g of sugar each time, but the scales went up in 2 g, so we could have been using 9 g or 11 g. We did the same number of stirs but it is hard to stir exactly the same each time.

Melissa used an everyday sense of fairness, which is not relevant here. Olivia used a good procedure for increasing the reliability of the experiment but this has nothing to do with fairness. Aiden's comment is about a fair test. Finley's is also about a fair test and attempts to quantify the fair testing. You will want students to progress from everyday ideas of fairness (Melissa's comment) to the idea of 'keep everything else the same' as in Aiden's comment (where the same person does the stirring) and then to a more critical evaluation of which factors are relevant to control (Finley's comment). They should also consider which factors cannot be controlled but must be managed by a sampling strategy, and have some quantitative notion of how much error is introduced by factors in the natural world that cannot be controlled. Examples of students' statements to criticise, to bring out these points, can be obtained either from examination board exemplar material or from your own students' work.

5.6 Teaching evaluation during whole scientific enquiries

There is a tendency in whole enquiries for teachers to try to teach all aspects of enquiry at the same time. This is not feasible. No teacher would try to teach all of secondary school chemistry in one lesson. Neither is it possible to teach all aspects of scientific enquiry through one scientific enquiry. It is important to have a clearly defined focus in terms of learning objectives, in this case learning about evaluation. Having taught isolated aspects of evaluation, students should practise applying these ideas in whole evaluations. These will need to be done in the context of a variety of kinds of enquiry as the focus for evaluation will be different in different kinds of enquiry.

Evaluation can be taught in the context of whole enquiries:

- when students are collecting their own data, as an integral part of the enquiry
- to focus on the relationship between data and theory and the difficulties in finding suitable evidence to support or refute theories
- by scaffolding learning using supports such as prompt sheets and writing frames
- using second-hand data (i.e. evaluating whole enquiries carried out by others).

5.7 Evaluation as an integral part of a whole investigation

Students find the evaluation of scientific enquiries difficult. For this reason they need support in learning how to evaluate. It should not be an activity that is done at home after all the data have been collected, but rather should be an integral part of the enquiry. As students plan their enquiries they need to think critically about how to obtain the best evidence from the outset and amend the plan to collect additional evidence as the enquiry proceeds. It is useful to encourage evaluation at various stages during an enquiry and to allocate class time for this. Students will then have time to learn how to evaluate and will be able to act on their evaluations and improve the quality of their enquiries.

◆ *Teaching suggestions*

1. Teacher–student and student–student interaction helps students to evaluate their work as they go along. You can ask them to consider questions like: 'Should we take more readings? How would that improve our evidence?'

   The example in Box 5.9 shows how the teacher wanted the students to consider whether they had a large enough sample, when they were carrying out a pattern-seeking enquiry designed to identify why some people could throw a tennis ball further than others.

Box 5.9 *Ongoing evaluation (adapted from* Targeted Learning*)*

Teacher:	'What have you found so far?'
Linda:	'John had the longest arm and did actually throw the furthest. I have the shortest arm and threw the least, with Jane as well.'
Teacher:	'But Jane has quite a long arm, so she doesn't fit the pattern well. And what about Cameron? Can you think of any way we could check out this idea more thoroughly.'
Paul:	'Try measuring a few more people.'
Linda:	'That would help. That way you would get more of a pattern.'
Teacher:	'It might or it might show that your prediction does not work.'

2. At the end of the enquiry you could ask your students to present their results to the rest of the class for scrutiny, justifying their results and conclusions against criticism. This helps them to focus on interpretation of evidence and possible weaknesses in evidence. To do this properly takes time and so this time must be planned into lessons. It also involves establishing a supportive atmosphere in the classroom. Students have to be taught how to criticise constructively.

 3. Examples of other students' reports of similar enquiries, whether from published sources (e.g. from AKSIS *Developing Understanding* Unit 20 or the QCA website) or

from your own school portfolios, can be criticised to practise applying appropriate judgements. Examples are:

- in a fair test: the control of variables and the range and number of readings
- in a classification: the sufficiency of examples and choice of characteristics
- in a survey: the sample size
- in developing a technique: the trial data.

5.8 Evaluating relationships between data and theory

Exploration of the relationships between evidence and explanatory models is particularly difficult for students and scientists alike. Students can be encouraged to evaluate whether evidence supports or refutes particular theories.

◆ *Teaching suggestions*

1. You could use the strategy of giving different students opposing hypotheses to test, for example 'magnesium combines with oxygen when it burns and so it gains weight' and 'when things burn phlogiston escapes from them so they weigh less'. Set your students the familiar investigation of burning magnesium, but unbeknown to them supply instruction sheets in two alternative versions, each stating one of the hypotheses. With different expectations of evidence, they must subsequently evaluate whether the evidence supports their theory. (Details are given in AKSIS *Developing Understanding*, Unit 21.)

2. You could carry out quick demonstrations of the behaviour of solids, liquids and gases and then ask students to offer explanations (see the QCA *Science: A Scheme of Work for Key Stage 3*, Unit 6G). They can proceed to discuss how convincing this evidence is in supporting a kinetic particle theory. Similarly they can discuss the evidence for a spherical earth and a heliocentric solar system, the germ theory of infection, continental drift, or any other topic, to evaluate how confident they are in the evidence. (See also pages 127–135, 143–149, 149–152.)

5.9 Using supports such as prompt sheets and writing frames to teach evaluation

 Prompt sheets and writing frames, and similar computer versions, can support the evaluation of scientific enquiries. They can be used to scaffold learning but you should make sure that they do not become merely a checklist of things for students to complete when doing an enquiry. (This is discussed, with examples, in Chapter 1, section 1.3.)

5.10 Teaching evaluation using second-hand data

When studying 'content' areas of science, there are opportunities to evaluate second-hand data from which the information is derived. It is also important for students to learn how to appraise critically the published reports of scientific work. This is a good opportunity to employ strategies of role play and writing for different audiences (see Chapter 1, page 9).

♦ *Teaching suggestions*

 1. You can ask students to scrutinise a media report of some current scientific issue, such as 'power lines cause cancer' to judge whether it contains sufficient evidence to convince us of the case it is arguing. Animal experimentation is a topic of interest to students, and literature produced from both sides of the argument can be scrutinised for the adequacy of its evidence. The internet can provide sources of evidence, but care must be taken in selecting sites which present reports at a level appropriate to secondary school students. SATIS units and the PRI resources listed in Chapter 1 'Other resources' contain examples written at an appropriate level for students. (See also pages 129–133, 169–172.)

2. When studying human growth (see QCA *Science: A Scheme of Work for Key Stage 3*, Unit 7B) you can give students some tables and graphs of the growth of boys and girls (the classic Tanner data is found in many textbooks). Help them to think about how many individuals are needed for measurement to ensure reliable information, and also what other factors should be considered. Discuss reasons for similarities and differences between the large data set and a school class set in terms of sample size and other factors. Be open to a range of ideas, for example the time and place in which the data were collected may make them atypical of contemporary populations.

5.11 Summary

The purpose of evaluation must not be lost in detailed activities. Throughout all the procedures and activities described students need to keep two key questions in mind:

1. Do I trust the evidence collected?
2. Does the evidence support the conclusions drawn and has effective use been made of all the evidence?

References

AKSIS Project *Developing Understanding* and *Getting to Grips with Graphs* (details are given in 'Other resources' in Chapter 1, page 16).

Goldsworthy, A., Watson, R. and Wood-Robinson, V. (2000) *Investigations: Targeted Learning*. ASE, Hatfield, Herts.

QCA (2000) *Science: A Scheme of Work for Key Stage 3*. Qualifications and Curriculum Authority, London.

The Science Investigator computer program, originally published by Essex Advisory and Inspection Services and refined in the AKSIS *Developing Understanding* pack (details are given in Chapter 1, page 16).

6 *Ideas and evidence: Introduction*

David Sang

In the chapters which make up the second half of this book, we look at some broader aspects of scientific enquiry, known as *ideas and evidence in science*. The skills and attitudes which students can develop in this area are related to those that they develop in investigative work. The two can support each other.

Our students, in their everyday lives, come across many science-related issues, for example a proposal to grow genetically modified maize in neighbouring fields, concerns about greenhouse gas emissions, and even the question of which new car the family might buy. How can they bring useful ideas about science to bear when considering such questions?

Our task is to help our students to understand more about the nature of scientific evidence (this is largely developed through their investigative work), and the ideas that scientists conjure up to explain the evidence they have collected.

6.1 The big ideas

Here is a general outline of the broad ideas that students can be expected to learn about during their secondary science course. They might look at these in the context of historical discoveries, or of current controversies in science and technology.

- Scientists report their work in refereed journals, that is their work is reviewed by their peers before publication.
- The public are informed about issues in science and technology through reports in the media. These reports have their limitations; they are often incomplete, biased, oversimplified, etc.
- Science may appear cold and objective, but this obscures the imaginative nature of theory making. Because scientific explanations are works of the imagination, it is likely that more than one explanation can be found for a given set of observations, and that more than one model can be devised as part of an explanation.

- Scientists are human, and they share the frailties of all humans. They may be reluctant to give up an old explanation when new evidence appears to contradict it. They may prefer one explanation over another because they are influenced by their own social, moral or religious views. Alternatively, they may support a new, alternative theory because they are anxious to dethrone the 'old guard'.
- New explanatory ideas are more likely to gain acceptance if they are the basis of predictions that can be tested and confirmed experimentally. (In their investigative work, students should have experienced the cycle of gathering evidence, devising an explanation and using their explanation to make testable predictions.)
- In practice, the real world is very complex. Scientists may find it difficult to untangle the complex causes of a particular outcome, particularly when the available evidence is limited, and when there is no well-established explanation linking causes and outcomes.
- Technology is based on science. It has produced many benefits, but it has also harmed people and the environment (even if unintentionally). We must try to weigh benefits against costs. At the same time, technology has often enabled science; many scientific advances have been made possible only by advances in technology.
- In the end, the decision to go ahead with a particular application of science may depend also on moral questions that may tip the costs-benefits balance.

6.2 Teaching approaches

Science teachers spend a large proportion of their time helping their students to get to grips with 'the facts' of science, that is the established knowledge and ideas of physics, chemistry and biology. Science may therefore appear to be a solid body of well-established knowledge. For some students, and for some teachers, there is satisfaction in knowing that they are on solid ground. What they need to know is in the textbooks; it can be learned, and reproduced in exams.

For some students, however, this is unsatisfactory. In other subjects, they learn to express their own thoughtful opinions. They learn also to assess evidence, and to think about the consequences of actions. These students may feel that science is dogmatic, ultimately hard-hearted and lacking in social concern.

Work on 'ideas and evidence' can show a different face of science: science as an ongoing, developing, human activity. Scientists form a community, with its own rules and ways of working. One day, some of our students will be the next generation of scientists; we can help them to see how they might have a role in that community.

The ideas we wish to convey require a different approach. We want students to be able to express and evaluate different ideas, and to consider how scientific ideas can be applied in a social and moral context. As teachers, we have to change our position; we are no longer the repository of the 'correct answer'. We have to help our students to devise their own questions, and to seek out and evaluate their own answers. The chapters that follow present a range of appropriate teaching strategies.

In Chapter 7, Mary Ratcliffe looks at how you can tackle some science-related issues, namely topics where there is a danger of generating more partisan heat than thoughtful light. She suggests some approaches that will allow students to stand back from controversial questions and to evaluate them.

In Chapter 8, Peter Ellis looks at some important points in scientific history and the lessons to be learned from them. Historical contexts often have the advantage that the science involved is more readily appreciated by students, and may form part of the curriculum they are studying.

In Chapter 9, Tony Sherborne suggests some further approaches to current issues, linked to topics within the curriculum.

Finally, in Chapter 10, Peter Campbell outlines the contribution science teachers can make to developing students' notions of citizenship. The ASE has developed a range of resources for teachers that will help them contribute to their school's obligations under the National Curriculum in England.

In these chapters, the authors have presented a range of teaching strategies. Each strategy is presented here in a specific context through the use of case studies, but each is general enough to be applied in a great range of contexts. (Some alternative contexts are suggested, with directions to appropriate resources.)

General resources

Ideas and Evidence in Science (2002) Folens, Dunstable, Beds. This pack with CD-ROM, covering a wide range of issues and controversies in science, both contemporary and historical, has been developed in conjunction with the ASE. Each unit contains a copiable student 'storyline' supported by colour and monochrome overhead transparencies; a teacher version of the storyline includes a series of points for discussion together with answers to the questions. Two of the units deal with the philosophy of science. There are also GCSE questions, answers and marking schemes.

There are 19 units:

The Human Genome project
Natural selection and Darwin
Fossils and evolution
Surrogate mothers
Tobacco and lung cancer
Causes of cancer; interpreting the data
Ozone depletion
Greenhouse effect
Are mobile phones safe?
Big Bang theory
Rutherford's model of the atom
Is there life elsewhere in the universe?
Ideas of Galileo
Alice Stewart and X-rays
Philosophy of science I
Philosophy of science II
GCSE questions, biology
GCSE questions, chemistry
GCSE questions, physics

Two sample pages from this pack are shown in Figure 6.1.

The Greenhouse Effect

Read the storyline in Student Worksheet 1. Working with others in your group, answer as many questions as you can.

A
1. What advantages will there be if the climate in Britain is much warmer than it is now?
2. What problems might there be in a warmer Britain?
3. How has the climate in Britain changed over the last 10 000 years?

B
1. Which fossil fuel provided energy for the Industrial Revolution?
2. Name three fossil fuels we use today, giving an advantage and disadvantage of each.
3. Identify two changes that might be linked to the enhanced greenhouse effect.

C
1. What alternative sources of power might reduce our dependence on fossil fuels? Why don't we use them now?
2. The concentration of carbon dioxide has been rising since the Industrial Revolution. In some years of the twentieth century the global average temperature did not rise, it fell. Does this show that the enhanced greenhouse effect must be wrong?
3. How might we reduce the amount of carbon dioxide in the Earth's atmosphere?

Extension questions

If global warming is being caused by an increase in the amount of carbon dioxide in the air, how can we reduce the problem? Here is one recent idea.

Oil and gas have been extracted from the rocks under the North Sea for many years. Since the oil and gas have been trapped there for millions of years, we know that the rocks don't leak. The idea is to pump carbon dioxide back into the reservoir rocks once all the valuable oil and gas have been removed.

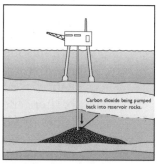

Carbon dioxide being pumped back into reservoir rocks.

Nobody knows if the idea will work or how the carbon dioxide might react with the underground rocks and possibly change them. This is likely to be a very expensive way to permanently remove carbon dioxide from the air. An alternative is to liquefy the carbon dioxide under pressure and dump it in the deep ocean. The liquefied gas is heavier than sea water and should stay at the bottom of the ocean. We hope.

1. How can the carbon dioxide be collected ready for disposal? Remember that air contains very little of it, about 0.03% by volume.

2. What are the dangers of storing unwanted carbon dioxide in these two ways?

3. Everyone on Earth will benefit if climate change can be halted or reversed. This means that everyone should pay towards the cost of removing carbon dioxide from the air. Do you agree?

4. **Research exercise**
   Think of a different way to reduce the concentration of carbon dioxide in the air. Find evidence from books, CDs or the web to support your ideas.

Ideas and Evidence in Science

Figure 6.1 *(continues opposite)*
Sample pages from Ideas and Evidence in Science *(2002), Folens Publishers, Dunstable.*

Answers ▷ *Ideas and Evidence in Science*

The Greenhouse Effect

 A

1. Lower fuel demands/less or lighter clothing required; 'better' climate/holiday at home not abroad; grow exotic plants, fruits and so on.

2. Conventional farming impossible, neither crops nor animals can withstand higher temperatures/new insects, parasites can survive causing new diseases; if sea level rises some areas will flood, such as London and East Anglia, and so on.

3. From an arctic landscape to a temperate climate/ almost no vegetation to the situation today/glaciers to rivers, and so on.

B

1. Coal was the main fossil fuel; found close to deposits of iron ore and limestone in Britain from which iron and steel are made.

2.

Fuel	Advantage	Disadvantage
Coal	Widely available in Britain, no need to import	Causes serious pollution when burned, solid is hard to transport
Oil	Liquid, easily transported, suitable for car engines	Limited local supplies, soon needs to be imported
Natural gas (methane)	'Clean' fuel, least polluting of the fossil fuels	Limited supplies, danger of explosion with gas-air

3. Melting of ice at the Poles/rising sea levels; changing weather patterns; movement of species to new areas; reduced fertility of soil in hotter areas; changes to forest cover.

C

1. Nuclear – problems with waste disposal.
   Solar thermal and solar photovoltaic – free fuel but systems themselves are expensive; no sunlight = no power.

2. Not necessarily, may be temporary variation in a rising trend.

3. Burn less fossil fuels; develop alternatives; plant forests; reduce energy demand; raise efficiency.

PRI *Ideas and Evidence Science Pack* (2001) Collins, London. This is another pack which contains a number of units; it has been developed by the Pupil Research Initiative (PRI), based at the Centre for Science Education, Sheffield Hallam University. Activities are presented in two forms, one of which is suitable for use in cover lessons. The text and graphics are presented in a customisable form, to allow users to integrate the units into their existing schemes of work. Visit the PRI website at www.shu.ac.uk/pri for further information.

Starter units

Which theory is correct?
Impossible? Maybe!
The uncertainty game
We have the technology
Newspaper headline bingo

Main units

Vaccines
Getting into the media
BSE and communicating science
How safe are mobile phones?
Death of the dinosaurs
Darwin and evolution
Wegener and continental drift
Understanding the universe
Dalton and particle theory
Cleaner energy
GM food
Climate on trial
The search for aliens
Predicting quakes and eruptions
Superbugs
Testing new drugs on people
Superhumans?
Cloning
Safe drinking water
Red giant
Polluted air
Crash tests
Speed cams
Energy drinks

Big Questions – the Nature of Scientific Enquiry, 2002, 4Learning (PO Box 100, Warwick CV34 6TZ, tel 01926 436444).

 A series of five television programmes in the *Science In Focus* series, presented by Adam Hart-Davies. These are available on one video.

Faraday's famous inventions
Charles Darwin's evolution
Mendel and the gene splicers
Mendeleyev's dream
Hubble's expanding universe

7 Ideas and evidence: Tackling science-related issues

Mary Ratcliffe

Most information about advances in scientific applications reaches us and our students through the media. There is frequently an emphasis on controversy, whether it be conflicting scientific evidence or the social and ethical implications. We form opinions about these issues based on our values and beliefs as well as any understanding of the underlying science.

Some students may feel hostile to science and technology. Others may feel that science should be expected to come up with solutions to social ills. For all of us, it is important that we develop strategies for approaching ethical dilemmas in science and technology. This allows students to appreciate science and technology as human endeavours, and it is important that future generations of scientists appreciate the ethical dimensions of scientific advancement. This chapter looks at several approaches to evaluating science-related issues. The teaching strategies suggested here will allow students to consider their opinions and values and open them to scrutiny in a relatively neutral and supportive context; they will also help students to identify the strengths and limitations of media reports on complex science issues.

7.1 Air pollution

This topic is suitable for use with younger secondary students. It allows them to look at:

- issues of the application of science, particularly the ethical dilemmas that result
- the positive and negative effects of scientific and technological developments on the environment and in other contexts
- how to take account of others' views, and why opinions may differ
- how scientists work today.

Context: Work on chemical reactions, particularly in relation to air pollution (see, for example, unit 9G in *Environmental Chemistry* of the Science Scheme of Work for Key Stage 3 in England; details in Chapter 1 'References', page 17.)

♦ *Learning outcomes*

Through this topic, students can learn:

- how to apply the notions of goals, rights and responsibilities in the context of an ethical dilemma
- that there is no single right answer to an ethical dilemma
- that different parties in ethical dilemmas will generally view the issue at hand in the context of their own interests.

♦ *Background*

This topic is dealt with using a strategy of 'goals, rights and responsibilities'. Why is this an appropriate approach?

Most applications of science and deliberations of science in the making have an impact on people, for example the differing scientific evidence for the safety of mobile phones (see section 9.2 on page 162), genetically modified foods and development of new health treatments. It is easy to generate an emotive discussion about issues such as testing the safety of consumer products on animals. Even scientifically based discussions and role plays can develop into a 'fight' to win rather than a reasoned debate. They may start out as a consideration of the scientific evidence and values in complex cases, but can become a personal campaign to make points by dominant individuals. The goals, rights and responsibilities strategy attempts to depersonalise controversy while at the same time considering an issue from the perspectives of different individuals or groups.

◆ *Teaching strategy*

Students are presented with a scenario or 'story' to which they can relate – either real or constructed for the purpose of highlighting particular issues. The scenario contains an ethical dilemma resulting from knowledge or application of science. The people in the story can either be individuals or identified groups. Students are then asked to consider the 'goals, rights and responsibilities' of the different people in the story. This is not role play – because students are not asked to put themselves in the person's shoes but to consider their goals, rights and responsibilities in relation to the particular issue. There is some empathy with the person but no attempt to model their actions.

Goals, rights and responsibilities may be in conflict in many situations. People may use all three to justify one course of action over another; usually one of them (i.e. a goal, a right or a responsibility) will be given as the most important reason. The intention of using this strategy is to expose the conflict to students and to consider how different legitimate decisions may be reached depending on the priorities. Once the scenario has been presented, the first part of the activity is to indicate what is meant by 'goals, rights and responsibilities'.

- **Goal:** This is what the person intends to accomplish through a particular action. A 'good' outcome might be regarded as morally correct regardless of how it was achieved – the idea of any means to an end.
- **Right:** This is the person's entitlement to a particular kind of treatment no matter what the consequences.
- **Responsibility:** This is the person's obligation to act or behave in a particular way.

Case study 1 which follows illustrates how these can be identified in a particular situation.

Case study 1: Sulphur dioxide levels and asthma

This fictional case study is based on data for a particular location (Middlesbrough). However, it can be adapted for different localities, people, etc.

 The DEFRA (Department for the Environment, Food and Rural Affairs) website contains details about air quality across the UK. Extensive information is available on the quantity of air pollutants in many UK towns and cities for past years and for the previous week. The web address is:

http://www.defra.gov.uk/environment/airquality/index.htm

The case study is based on data for sulphur dioxide levels recorded for 1997. These data are shown in Box 7.1 as information for students.

Box 7.1 *Information for students*

How should we deal with pollution by acidic gases?

Monitoring equipment records sulphur dioxide levels in many towns, cities and rural locations in the UK. One monitor can be found on a housing estate near a further education college. The area has some local industry around it and much traffic uses the roads in the area.

In 1997 the sulphur dioxide pollution here was one of the highest in the UK. Although most of the year the levels were low on four days there was moderate pollution, on one day high pollution and on one day very high pollution.

Helen is a mother of two children: Anne, 12, who is asthmatic and Paul, 16, who has just finished his GCSEs and is expecting to go to the further education college in September. She works as a solicitor in a nearby town centre. Mike, her husband, works in a local chemical works, which contributes some sulphur dioxide pollution. Both Mike and Helen drive large cars and use them on most journeys.

The local council is aware of the sulphur dioxide pollution levels and is considering the following measures to reduce pollution and energy consumption.

- It wants to raise car park charges in the town centre and stop all-day parking there.
- It intends to help enforce the law on promoting low pollution levels from the factory. This could lead to a reduction in jobs in the chemical works.

Anne wants the family to move to a rural area and get rid of one of their cars. Should the family follow Anne's wishes?

Identifying goals, rights and responsibilities

Students may have an immediate reaction to the 'should' question. The intention of the goals, rights and responsibilities strategy is to look at the case from a number of angles, picking up conflicts.

Each small group within the class considers the goals, rights and responsibilities of one particular person within this scenario. This means that, on average, each person might be considered by two groups. This can take about 10–15 minutes. Groups can then be asked to consider a different person, if desired or time allows. The scenario could be extended to include the boss of the chemical factory and the leader of the council.

Students may find they want additional information about the family. The scenario can obviously be expanded. Alternatively a real, local problem can be considered using the same idea of objective analysis rather than role play.

You can then collect together the class views in a table, which might look something like Table 7.1.

Table 7.1

Family member	Goals	Rights	Responsibilities
Anne	To be healthy and cope with her asthma	To have her own opinion	To go with her parents' wishes
	To encourage the family to be environmentally conscious	To have her asthma treated	
Paul	To further his education	To have access to relevant education opportunities	To respect the views of his sister and parents
	Not to be exposed to too much pollution	To leave home if necessary	
Helen	To continue in a good job	To earn a living	To limit pollution
	To be independent in travelling		To promote a happy family
Mike	To continue in local employment	To earn a living	To limit pollution
		To be respected by his children	To obey the law

The statements in the table are opinions and these may vary depending on the perspective of the discussion groups, but nevertheless they show the conflict between different people when considering issues of reducing pollution. The intention of collecting the views together in this way is to:

- show the conflict
- show that it is not easy to resolve the conflict
- show that different solutions may be reasonable depending on the balance between goals, rights and responsibilities of different people.

The aim of the activity is not to reach a decision but to open up to scrutiny the arguments that students might present. Students may want to have their say in terms of making the decision, but the framework allows the teacher to encourage and expect reasoned arguments for a particular decision based on balancing different values and considering the scientific evidence. This activity can be followed by asking each group to present arguments for and against the change proposed (in this case the family following Anne's wishes) and, importantly, identify the scientific evidence and values used in mounting these arguments.

◆ *Other sources*

For another case study, see the ASE's *Citizenship through Science* materials (details at the end of Chapter 10, page 177).

7.2 The use of biomass as fuel

This topic is suitable for use with younger secondary students. It allows them to:

- address issues of the application of science
- think about the positive and negative effects of scientific and technological developments on the environment and in other contexts
- take account of others' views and understand why opinions may differ.

Context: Work on the use of energy resources.

◆ *Learning outcomes*

Through this topic, students can learn:

- to present an argument or evaluate an argument in which advantages and disadvantages of a particular course of action are considered
- to distinguish between scientific evidence and value judgements.

◆ *Background*

Students value discussion in science lessons, particularly where they can express and compare opinions about science in the news or contentious science that impinges on their daily lives. However, as science teachers, we can find controversy and student discussion difficult to handle for a number of reasons:

- the outcomes of a discussion may appear uncertain
- students (and teachers) may have very different views, based on a wide variety of belief systems
- the scientific evidence may be inconclusive and information about the issue may be incomplete
- students' (and teachers'?) expectations of science lessons are that the outcome is something certain, that is scientific knowledge that has stood the test of time.

The teaching strategy suggested is intended to realise the benefits of student discussion in small groups within a clear structure.

◆ *Teaching strategy*

Case study 2 shown right illustrates the use of *consequence mapping*. For the teacher, the steps in its implementation are:

1. Identify a good 'what if...' question. (This is the most creative and tricky step.)
2. Identify the opposing 'what if...' question.
3. Have half the groups in the class work on one question, and half on the other.
4. Ask each small group (three or four students) to produce a consequence map.

Once students have developed their maps, they can identify:

- positive and negative consequences (they will probably realise that this depends on the values of the individual or group)
- consequences that are certain and those that are uncertain
- statements that are fact (scientific or otherwise) and those that are opinion.

This strategy should bring structure to the small group discussion, allowing all students to contribute.

Case study 2: Comparing the use of biomass energy in different countries

The current target in the UK is to have 10% of electricity generated from renewable fuels by 2010, subject to the cost to consumers being acceptable. In 1999, just under 3% of electricity was generated from renewable fuels, with hydroelectricity forming about half of this contribution. Most biomass fuel is currently used for heating.

Students can be presented with some basic information about biomass as a fuel from one of the websites below (numbers 1 and 2) or from available texts. For example, fast-growing trees such as willow can be grown, harvested and burnt for heating or the generation of electricity. In other countries, such as Brazil, sugar cane can be harvested, fermented to produce ethanol and the ethanol used as a fuel, including for transport. Plant and animal waste can be rotted to allow generation of biogas (methane).

The deliberate growing of crops for biomass energy in the US and UK can be contrasted with energy use in Africa (website number 3). 'Biomass accounts for as much as two-thirds of total African final energy consumption. In comparison, biomass accounts for about 3% of final energy consumption in OECD countries. Deforestation is now one of the most pressing environmental problems faced by most African nations, and one of the primary causes of deforestation is wood utilisation for fuel.'

Useful websites
1. http://www.dti.gov.uk/renewable/ed_pack/index.html
   On the Department of Trade and Industry website – an education pack on renewable energy. The main DTI website (http://www.dti.gov.uk) has detailed information on energy sources.
2. http://www.britishbiogen.co.uk/
   A useful source of information about the British bioenergy industry.
3. http://www.newafrica.com/energy/
   Some basic statistics on energy use in Africa, including biomass.
4. http://www.shell.com/rw-br
   Developing commercial opportunities in biomass, solar, wind and forestry as energy sources.
5. http://www.defra.gov.uk/
   Department of Environment, Food and Rural Affairs website, with useful information including a schools page containing resources on global warming:
   http://www.defra.gov.uk/environment/climatechange/schools/teachers/index.html

Identifying consequences

After students have gained some basic information about biomass energy, they can produce consequence maps. A pair of questions is posed to the class; half of the students work on one, half on the other. They should produce their maps on A3 sheets, as posters or OHPs.

Some possible pairs of questions are given in Box 7.2; students need select only one pair. An example of a consequence map for question B1 is shown below.

Box 7.2 *Questions for consequence maps*

> **A1:** What if the target '10% of electricity from renewable fuels' all had to come from biomass fuels?
>
> **A2:** What if the target '10% of electricity from renewable fuels' was for other renewable fuels and did not include biomass fuels?
>
> **B1:** What if the use of biomass fuels had to increase a lot?
>
> **B2:** What if the use of biomass fuels were stopped?
>
> **C1:** What if the local houses or school had to be heated by biomass fuel from next year?
>
> **C2:** What if the local houses or school were not allowed to use renewable fuels, such as biomass, for heating?

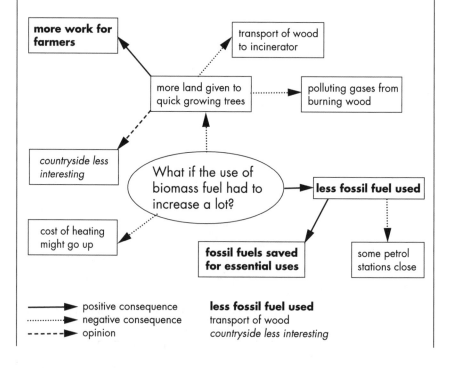

There are a number of ways of drawing the activity together, depending on the particular learning outcome emphasised:

- Students can display their maps around the room. They can then be asked to identify points for using biomass as a fuel, and points against.
- Students can be given someone else's map and asked to identify what they agree with, and what they disagree with.
- Students can be asked to highlight the scientific evidence for or against increasing the use of biomass fuels.

The teacher can collect on the board/OHP information from each group under the headings 'positive consequences of using biomass fuels' and 'positive consequences of not using biomass fuels'. Although this could be done without producing consequence maps, the small group discussion involved in producing the map allows students to tease out the issues for themselves and compare ideas.

It is likely that the following general points will arise, as shown in the sample map, and are worth emphasising:

- There are both positive and negative consequences to any action.
- Whether consequences are positive or negative depends on the point of view of different groups in society (e.g. some people will be pleased to see petrol stations close, while others will be unhappy).
- The science is embedded in the map rather than being explicit – value judgements are to the fore.
- Many aspects are uncertain – sometimes this is because of lack of information in the class, and often the information is incomplete because complex models are needed to take account of all the different possibilities and their effect on each other. Nevertheless we develop opinions from incomplete information.

◆ *Reflections on consequence mapping*

The 'consequence-mapping' strategy itself has advantages and disadvantages.

Advantages
- The idea behind consequence mapping is easy to understand and accessible and interesting to most students.
- Consequence maps can be used for a wide range of issues.
- Consequence maps provide a visual representation of the issues, which is useful for display.
- The technique encourages open thinking.

Disadvantages
- Identification of the most appropriate 'What if. . .' question requires careful thought.
- Students can record factors rather than consequences and may focus on only initial rather than secondary consequences.
- The method does not always encourage students to focus on key consequences.
- Mapping usually requires some further activity to evaluate the consequences, as shown above.

◆ *Other sources*

You can find out more about the technique of consequence mapping, together with further examples, in the ASE's *Citizenship through Science* materials (details at the end of Chapter 10), and in *Teaching Ethical Aspects of Science*, by Patrick Fullick & Mary Ratcliffe (1996) Bassett Press, Southampton.

7.3 Using media reports

This topic is suitable for use with older secondary students. It allows them to look at:

- how scientific ideas are presented, evaluated and disseminated
- the power and limitations of science in addressing industrial, social and environmental questions – including the kinds of questions science can and cannot answer, uncertainties in scientific knowledge, and the ethical issues involved.

Context: Work on ecology, particularly the carbon cycle.

◆ *Learning outcomes*

Through this topic, students can learn:

- to distinguish between scientific evidence and value judgements
- to interpret media reports of science, in considering the nature of the scientific evidence presented and its certainty.

♦ *Background*

We, and our students, get much of the information about science issues that impact on our lives from media reports. These reports, whether they are in newspapers or magazines, or on television or the internet, often contain some information about the science behind the issue and comments on the social impact of the science. Those issues that have national impact are normally complex for two reasons:

1. The scientific evidence is not clear cut: the data, evidence and implications can be different depending on which scientists are researching the issue, the models and theories they use to interpret data and the wider acceptance of the evidence within the scientific community.
2. The social impact of cutting-edge science can be difficult to disentangle from the controversies about the scientific evidence and often brings into consideration issues of ethics, economics, politics, the environment, etc.

The intention of this unit is to enable students to examine media reports of scientific issues with a view to considering the nature and provenance of the scientific evidence and weighing evidence alongside value judgements.

♦ *Teaching strategy*

Any information about controversial issues in the media at the time of writing is likely to be out of date by the time you read this book. This is one of the considerations facing the integration of 'ideas and evidence' into science lessons. The strategy adopted in Case study 3 is, therefore, to illustrate the use of media reports with an issue from upper secondary science, emphasising possible sources of current information.

Case study 3: Using media reports: greenhouse gases

Media reports can be brought in by students or selected by the teacher from magazines such as *New Scientist* or *Focus* (a colourful monthly magazine on current science and technology issues aimed at the general public). However, there are copyright issues in photocopying from some newspaper and magazine sources. You need to check the conditions of the photocopying licence in your institution. Up-to-date

ICT reports that show some of the nature of scientific research can be found on some websites, such as the following:

---

http://www.edit.legend.yorks.com/sciesec.html

This site provides an annotated list of many useful secondary science websites.

http://whyfiles.org/

This US site shows 'the science behind the news'. It provides a new story each week focusing on an area of current interest or concern. The science and the ongoing research are explained in a lively manner. Pages can be downloaded for use as resources as indicated below. Alternatively, students can search this site for relevant articles as part of their own research (see section 7.4, page 133).

One problem busy teachers face is finding the time to search for relevant articles in the media. With the increased emphasis on literacy in science, why not encourage your school library to subscribe to *Focus* or *New Scientist* and make it a science club project for some students to keep abreast of issues in the press? If you subscribe to the *New Scientist* newsletter on their website, www.newscientist.com, you will receive an email each week showing the relevant updates.

Another source of articles is *Catalyst*, the GCSE science review. It regularly carries 'Life in science' articles in which practising scientists describe their work. Many of the other feature articles relate current research findings to GCSE science topics.

Once you or students have media reports, how do you use them to best effect? Clearly this depends on the intended learning outcomes. If the intention is for students to appreciate that the topic they are studying is of current media and research interest then reports can be provided as background reading.

However, the emphasis in this case study is on using reports to illustrate features of the conduct of science research, notably:

- who conducts scientific research
- how scientists collect evidence
- how scientists go beyond the evidence in constructing explanatory theories and models
- the process of peer review
- the limitations of scientific research.

Looking at news reports

These features can be brought out by considering the news reports shown below. You may consider that some of the language in these reports is beyond the level of some of your students. However, class reading followed by identification of problems with terminology can allow most students to get the thrust of the report.

Time bomb of UK wetlands

As global warming continues, Britain's bogs could release tonnes of the greenhouse gas carbon dioxide (CO_2), according to Chris Freeman from the University of Wales, Bangor.

We humans are polluting the world by burning fossil fuels. But carbon dioxide is locked away in peat bogs, because here plants and animals decay very slowly and so do not release their carbon dioxide back into the atmosphere. These bogs now contain almost 30% of the world's soil carbon.

But Freeman is worried. Carbon is stored out of harm's way because a single enzyme, phenol oxidase, does not work in the oxygen-starved conditions of a bog. If droughts caused by global warming dried out the peat bogs, this enzyme would be able to work, and atmospheric carbon dioxide levels could double.

Fragile balance: 'Phenol oxidase could be a fragile "latch" mechanism holding in place a vast carbon store of 455 gigatonnes,' says Freeman. 'It's a reminder that we need to protect our wetland environments.'

On a more positive note, however, US researchers say the Earth may be fighting back. They studied a patch of grassland over five years and found that when it was exposed to excess carbon dioxide, the balance of microbes and plants shifted to allow more carbon dioxide to be locked away below ground. So there could be a safety valve – at least for now.

From *Focus* (2001) **101** (April), p. 27

Forget rainforests
Fred Pearce

Rising levels of greenhouse gases have led to faster tree growth in arid regions. The discovery boosts the case for planting forests in dry areas to combat the effects of global warming. Plants combine water with carbon dioxide to create complex chemicals. Xiahong Feng of Dartmouth College in Hanover, New Hampshire, has shown that the rise in atmospheric carbon dioxide over the past 200 years has made this process more efficient.

Feng measured the changing ratio of different isotopes of carbon in the annual growth rings of a range of American trees. Short-term fluctuations reflected seasonal weather patterns, but Feng detected an underlying trend that matched the rise in global carbon dioxide levels (Geochimica et Cosmochimica Acta, vol 63, p. 1891).

'The rate of increase [in water-use efficiency] started low in the 19th century, but increased rapidly for most trees in the 20th century', she says.

Water use is not the only factor that determines how fast plants grow. The availability of nutrients and competition from other plants also affect growth. But Feng says that in arid environments, where moisture limits tree growth, biomass may have increased as a result.

This finding may help climatologists balance the Earth's carbon budget. Around half the carbon dioxide in the atmosphere that arises from human activity swiftly disappears. The oceans absorb some of it, but most researchers believe that much of the rest is absorbed by forests in cooler temperate regions, including Europe and North America. Feng's results, however, suggest that forests in arid regions may be more important than anyone realised.

The study will also encourage a group of hydrologists calling for an international effort to plant trees in the dry areas of the world, which are mostly covered by scrub and grazed by cattle or wildlife. After a meeting at UNESCO's headquarters in Paris in December, a group chaired by Arie Issar of the Jacob Blaustein Institute for Desert Research at Ben-Gurion University of the Negev in Israel proposed replanting vegetation in grazed areas to create vast parklands.

The countries that signed the UN Climate Change Convention have discussed replanting tropical rainforests, but Issar argues that the demands of agriculture and urbanisation in those regions threaten this approach. In contrast, in most dry lands, the density of the population is low and the demand for land insubstantial, he says.

There is often enough water in deserts for growing trees, according to Issar, though this may mean reviving ancient irrigation systems and tapping underground water. Although forests in dry lands grow at between a tenth and a quarter of the rate of tropical rainforest, he argues that the huge amounts of land available for tree planting in many arid countries should be ignored in the race to halt global warming. Feng's discovery may tip the economic balance towards planting dryland forests.

From *New Scientist* (1999) 2 October

Posing questions

To guide students' reading of media reports such as those shown above, it is a great help to provide them with a series of questions. Responses to these can be discussed in small groups, or you can ask your students to provide written answers.

Six questions are given below, together with some points for each that you might draw out in discussions of students' answers. The key ideas relate to the way scientists work and the implications of scientific research.

These questions can also help to develop critical skills in evaluating evidence. Questions 2–5 are related to similar questions that students should be asking when they are evaluating their own investigations.

1. **Who has done the research?** In both articles, university researchers are indicated. This fact can be used to show the location of much science research. Although these articles do not emphasise research teams, many others show that groups of scientists have been involved in both the research and the peer review.

2. **What is their conclusion from the research?** This is normally shown in the first few sentences, as illustrated in these articles.

3. **What evidence do they have for their conclusion?** This can be the point at which students realise the incompleteness of details presented in media reports. They should consider:

   • Is there information about what was observed or measured (or both)?
   • Are there data to show these observations or measurements?

   New Scientist articles normally give some indication of the collection and nature of the data. This won't be enough for students to evaluate the data, nor are they likely to have sufficient expertise in the subject to evaluate the original research. However, good media reports will indicate the basis for conclusions.

4. **What theories or models are they using to explain their evidence?** The purpose of this question is to help students realise that findings don't just emerge from data. Scientists develop hypotheses or refine existing models as they undertake data collection and analysis. For example, both articles assume the accepted mechanism for photosynthesis and are testing out ideas about carbon dioxide balance and global warming.

5. **How certain are the scientists of their conclusions? Do other scientists agree with their findings?** Again it is easier to judge this in some reports more than in others. The conditional

tense in the *Focus* article implies that the conclusion is not absolute, but not enough data are provided to judge further. The *New Scientist* article shows that the original research was reported in a peer-reviewed journal. This can be emphasised as showing that other scientists consider that the research was carried out effectively and reasonable conclusions were drawn. The certainty of the findings is open to others to attempt to replicate further.

Some reports will have quotes from researchers who take a sceptical view of the research findings, illustrating the difficulty of being certain from one study. Some of this different perspective on evidence from one study is shown in the latter part of the *Focus* article.

6. **How should we act on the research findings?** This question is posed to enable students to consider where and how value judgements are used in dealing with the outcomes of scientific research. It is at this point that strategies for dealing with the consideration of scientific evidence alongside value judgements can be reinforced.

♦ *Taking it further*

Once students have studied one or two reports, you have the opportunity to take things further so that they make use of what they have learnt. For example, from the two reports above, the following activities could be undertaken:

♦ **Goals, rights and responsibilities** – see page 118. Should peat bogs be dug up to provide peat for gardens, fuel for fires?

People involved include: the peat bog owner, the environmental campaigner concerned about the peat bog as a habitat for endangered species, the home owner in the local community who wishes to use peat as a fuel, and the gardener in another part of the country who wants a good source of peat for the garden.

♦ **Consequence mapping** – see page 122. Possible 'what if?' questions include:

- What if large forests were grown in arid regions?
- What if governments of arid regions banned the growing of trees there?
- What if peat bogs were left untouched but dried out?

♦ **Considering risks and benefits.** Produce a structured table showing the advantages and disadvantages of:

- relying on growing forests in arid regions to reduce global warming
- harvesting peat from peat bogs
- replanting tropical rainforests.

♦ **Writing a media report.** Finally, students can present the findings of their own investigations in the form of a media report to reinforce the nature of media reporting. This comparison may allow students to appreciate the limit to the detail in media reporting and thus some of the issues of making sense of media reports.

♦ *Further reading*

Fullick, P. L. & Ratcliffe, M. (eds) (1996) *Teaching Ethical Aspects of Science*. Bassett Press, Southampton. This book gives a detailed rationale behind the strategies described here, and outlines additional case studies suitable for upper secondary school students. It is available from ASE Booksales.

Ratcliffe, M. (2001) Teaching for understanding in scientific enquiry, pp. 28–46. In Amos, S. & Boohan, R. (eds) *Aspects of Teaching Secondary Science: Perspectives in Practice*. RoutledgeFalmer, London. This chapter gives further details of the rationale and effect of using evaluation of media reports with science classes.

7.4 Researching a sensitive issue

This approach is suitable for use with older secondary students. It allows them to:

- consider the power and limitations of science in addressing industrial, social and environmental questions (including the kinds of questions science can and cannot answer); uncertainties in scientific knowledge; and the ethical issues involved.

Context: Any contemporary science issue.

♦ *Learning outcomes*

Through this approach, students can learn to:

- research a sensitive issue and distinguish the different value positions people might adopt
- use a method of weighing advantages and disadvantages.

♦ *Background*

Younger secondary students have been introduced to the complexity of socio-scientific issues. They have started to consider the benefits and risks or advantages and disadvantages of the social application of science and technology. This unit is intended to allow students to develop their understanding of how to weigh evidence and consider risks and benefits, particularly when considering the ethics of particular stances. The unit should also encourage research skills and independent learning.

The skills of searching and evaluating evidence, weighing risks and benefits and developing a thoughtful opinion are all important for citizenship (see Chapter 10).

♦ *Teaching strategy*

This activity is designed to build on the strategies used previously (consequence mapping; goals, rights and responsibilities; and cost-benefit analysis). Rather than simply being provided with information, students are now encouraged to find or evaluate information sources (or both). For this to be effective, they should be given a clear structure within which to work, as in Case study 4.

Case study 4: Structuring independent research

1. Identifying, negotiating or allowing students to identify a particular controversial or sensitive issue

Students may have more stake in undertaking the research and seeing the project through if they have been involved in identifying the issue. However, you (and your students) may need to clarify the issue so that its scope is appropriate in terms of information that is likely to be available.

Some questions that can be relevant to the science content of GCSE courses include:

- Should hormones be used to control the fertility of animals?
- Should hormones be used to promote plant growth?
- Should cloning of animals be encouraged?
- Should genetic engineering of plants to promote disease resistance be encouraged?

These issues require a consideration of the underlying scientific evidence, evaluation of likelihood of outcomes from any actions and consideration of different value positions.

---

2. Researching the issue

Before considering and presenting arguments about the issue, students can clarify their understanding of the science involved and the nature of any scientific research – particularly if controversy and conflicting views on 'new' science exist.

ICT This part allows students to use their research skills, assuming that these have been taught. In particular, students can do a web search using keywords in a structured way. (For further information on web searching see *How to Manage Pupil Research Using the www* – ASE citizenship materials, see end of Chapter 10 for details.)

Alternatively, students can compare information presented in media reports with that from science reference books or textbooks.

Students can present a summary of their research showing:

- what is well-established scientific knowledge about the issue
- what is contested knowledge
- what is unknown.

3. Evaluation of the issue

For this part of the activity, students can be encouraged to use one of the strategies outlined earlier for considering a controversial issue (e.g. consequence mapping; goals, rights and responsibilities; analysis of media reports; cost-benefit analysis). But this time they are using it independently – that is, this project will be able to give a measure of their progress in ethical thinking.

They can add to their research by conducting a survey of fellow students or people in the local community, to gain a perspective on other people's views. Alternatively they can collect newspaper reports and see what value positions are presented in the media. They should be encouraged to recognise and identify the diversity of opinions that may emerge about the issue.

Students should be encouraged to present informed arguments. They can argue for or against an issue or present a balanced argument, but if they wish to argue on a particular side they should be able to acknowledge counter-arguments. They should be able to build on their earlier experience of considering value positions.

Ideas and evidence: Learning from history

Peter Ellis

The history of science and technology has a place in teaching science as it connects the content of a science course with people who are scientists or who use science. It also reveals science to be an ever-changing, uncertain and, most importantly, a human endeavour.

Through historical case studies, your students can tackle all aspects of ideas and evidence. This should not be seen as something additional to the teaching of scientific concepts; rather, by building historical examples into your teaching, the historical treatment of ideas can reinforce your students' understanding of scientific reasoning.

Various teaching strategies can be used to deliver the ideas and evidence statements through historical examples. The following sections give examples of student activities, historical contexts and possible issues. You can follow these examples directly, or use them as models to develop your own case studies.

8.1 William Withering and digitalis

This topic is suitable for use with younger secondary students. It allows them to look at:

- the interplay between empirical questions and evidence
- the importance of testing explanations and making predictions
- the way scientists worked in the past.

Context: Work on humans as organisms, particularly work on the blood system and health.

◆ *Learning outcomes*

Through this topic, students can learn that:

- scientists make observations and develop ideas
- scientists test ideas and predictions by experiment or trials
- in the past (the eighteenth century) scientists usually worked on their own, but communicated with other interested people
- in the past, medicines could be tested on patients, unlike today when new drugs require rigorous, controlled trials.

◆ *Background*

In the eighteenth century understanding about the role of the heart in the circulation of the blood was growing but there was little that doctors could do to treat the symptoms now known to result from heart disease. Traditional treatments such as blood letting and herbal remedies were generally ineffective.

William Withering was in many ways a typical doctor – better than most as shown by his success in practice in Birmingham – but he was trained in traditional methods. Where he was different from others was in following up a report of an old woman who seemed to have some success in treating the 'dropsy', the name at that time for the symptoms of heart disease. His knowledge of botany (encouraged by his wife) and natural medicines (he was the son of an apothecary) helped him to pick out the active ingredient, foxglove, which was known to be toxic. His scientific ability is shown in the way that he tackled the following questions:

- Can foxglove be used to alleviate the symptoms of dropsy without killing the patient?
- Which parts of the plant contain the active material and is it produced throughout the year?
- What is the best form in which to administer the drug?
- How much of the drug produces the desired effect and how frequently should it be administered?

Withering collected evidence and tested predictions by prescribing foxglove extract to over 160 of his poor patients over a ten-year period. Again, his scientific skills are shown in the detailed reports he maintained on each of his case studies. He was honest in reporting failures as well as successes and after ten years was confident about stating the quantity and form of the extract that would produce beneficial results in specific cases. He reported his findings to the Medical Society of Edinburgh, the Lunar Society in Birmingham, the Royal Society in London and to his medical colleagues through his book *An Account of the Foxglove*.

There were, of course, limitations to the work. Withering offered no explanation for the action of foxglove, nor did he do any chemical analysis of the active ingredient. Nevertheless his work is a good example of the practice of science and his findings had lasting value – indeed, an extract of foxglove (digitalis) and its synthetic derivatives are still used in the treatment of heart disease. Ethical questions are also raised by his use of unwitting patients for his experiments. At least his patients were suffering from various symptoms for which there were no other known cures – unlike Jenner's use of a perfectly fit boy to test his vaccination for smallpox.

◆ *Teaching strategies*

The story of William Withering can be found in a number of sources (see below). This story can be recounted to students or used in a text-based exercise. Box 8.1 gives an outline of the story for students.

Box 8.1 *Background information for students on William Withering*

William Withering and digitalis

In 1775 William Withering, aged 34, became doctor to the wealthy people of Birmingham but he set aside an hour a day to treat poor people. The same year he heard about an old lady who was said to be able to cure the 'dropsy'. Dropsy was a common ailment in which patients suffered from shortness of breath, and became bloated because water was trapped in body cavities. Many people eventually died from the illness, which is now known to be heart disease. The old lady's remedy was a mixture of various plant materials but Withering recognised one, extract of foxglove, as possibly having some effect. Foxglove was a known poison. He began a long series of tests in which he used different parts of the plant, which were dried and sometimes ground to a powder.

Withering began to give the foxglove to his poor patients who suffered from dropsy. He varied the amounts, always wary of giving a fatal dose, and tried different methods of giving the medicine (e.g. in a drink of tea, as a tablet or powder, or boiled in water to form a 'decoction'). For each patient he recorded the symptoms, the amount and form of the dose and the effects. After ten years he wrote up his work and the results from 163 patients in a book called *An Account of the Foxglove*.

By this time Withering was a rich and respected doctor. He attended meetings of the Lunar Society: a group of men interested in science. People listened to his ideas and other doctors copied his remedy. Extract of foxglove, or 'digitalis', became an accepted treatment for the dropsy and is still used today in the treatment of some forms of heart disease.

The story is also suitable for treatment by role play and drama. The students' notes in Box 8.2 give brief details of the main characters in the story. Students can be given the roles and asked to improvise the story, or they may wish to put together a scripted version that can be performed to other groups of students or recorded on video tape. Other possibilities are recording a 'radio' play or a documentary version in which the characters are interviewed.

Box 8.2 *Characters in the William Withering story*

Characters for drama

William Withering (1741–1799): Born in Shropshire, the son of an apothecary (someone who prepares and sells herbal medicines). Studies medicine in Edinburgh and starts work in Stafford in 1767 but earns only £100 a year, not a lot. Marries Helena Cookes in 1772 and becomes interested in botany. Meets Erasmus Darwin who recommends him to take over the medical practice in Birmingham in 1775. Is encouraged by Helena to collect plants on his journeys. Begins to suffer from the disease, tuberculosis (which affects the lungs and breathing).

1775 Begins trials of foxglove extract.
1776 Publishes a book on British plants. Becoming rich, he buys land in central Birmingham.
1785 Publishes *An Account of the Foxglove* and joins the Royal Society.
1788 Argues with Robert Darwin (son of Erasmus) over treatment of a patient. Health becoming worse, he spends the winters of 1792 and 1793 in Portugal (for the clean, warm air).
Prudish, fussy, a bit of a bore and irritable as he gets older. Concerned for his patients but is unhappy if people disagree with his views.

Helena Withering: Daughter of a lawyer in Stafford. One of Withering's first patients, marries him in 1772 and is much younger than him. A keen artist, particularly of flowers, she encourages Withering to find samples for her to paint. Plays the role of supportive wife. Large house with servants to manage. Entertains Withering's friends: other doctors, and members of the Lunar Society such as Erasmus Darwin.

Old lady: Living in the country in Shropshire. Probably known as a witch to some of her neighbours. Knows the recipes for many old remedies for ailments. Her remedy for dropsy contains about 20 different herbs as well as foxglove. The mixture causes patients to be violently sick but does improve their symptoms. Happily gives Withering her recipe.

Erasmus Darwin [grandfather of Charles Darwin]: Elderly founder member of the Lunar Society. Well-known authority on scientific matters and knows all the famous scientists and many rich factory owners and land owners. Recommends Withering to take over the Birmingham medical practice of old Dr Small. Advises Withering on publication of his book on British plants.

Dr Stokes: Another doctor and friend of Withering. Reports his work to the Medical Society of Edinburgh, where it is well-received.

Patient H: Aged 60, near death, fond of drink. Very ill with dropsy (shortness of breath, bloated feeling). Has heard about Withering's remedy and wants to try it. Given digitalis but no effect.

Patient S: Aged 49, suffers from asthma as well as dropsy. Told to take some powdered foxglove every 2 hours until there is an effect. Soon much improved.

Patient G: Aged 86. A cheerful, sensible old man who has always looked after his health. Suffers from asthma as well as dropsy and for some years has lost his appetite for meat. Given some pills of foxglove, soon produces a lot of urine and his swellings go down. He feels restored to health.

Patient C: Aged 58. Has suffered from dropsy for some time. Was treated with foxglove with success for a year. Now has no appetite, is very weak and turning from yellow to black. Is treated with digitalis but with no effect and dies soon after.

Other characters: Relatives of patients; members of Lunar Society and Royal Society, such as Joseph Priestley, James Watt and Josiah Wedgwood, who hear Withering talk about his work; servants.

Suggested scenes

1. With Erasmus Darwin, William and Helena Withering discuss botany, as well as their possible move to Birmingham – because of the opportunity for a good career; concern for patients; concern about lack of treatment for symptoms of dropsy (e.g. shortness of breath, bloated feeling, lack of appetite).
2. Withering visits old lady, discusses the remedy, leaves with recipe.
3. Withering and Helena discuss the use of foxglove: which parts of the plant? when should it be collected (spring, summer, autumn)? how should it be used: dried as a powder or pressed into tablets, boiled up with water to make a 'decoction', or mixed with hot water to make an 'infusion' (like tea)? worried about the dose: too much and the patient is poisoned.
4. Tries out the remedy on some patients: a decoction is no use, but tablets, powders and infusion of leaves seem to work.
5. Withering and Stokes discuss the results of the study.
6. Withering is congratulated by members of the Royal Society after reporting his work.

◆ *Other sources*

Ellis, P. (ed.) (1999) *breakthrough*. Photocopiable pupils' materials published by PRE*text* (see Other resources, page 155).

Hart-Davies, A. & Bader, P. (1997) *The Local Heroes Book of British Ingenuity*. Sutton, Stroud.

Millar, D., Millar, I., Millar, J. & Millar, M. (2002) *The Cambridge Dictionary of Scientists*. CUP, Cambridge.

Whitmore Peck, T. & Douglas Wilkinson, K. (1950) *William Withering of Birmingham*. J. Wright, Bristol (includes facsimile of *An Account of the Foxglove*).

8.2 Hans Christian Oersted and electromagnetism

This topic is suitable for use with both younger and older secondary students. Younger students can look at:

- Oersted's search for evidence to back up his ideas
- Oersted's career as a public lecturer and amateur researcher
- the role of experimentation in his work.

Older students might focus on:

- how public lectures were (and still are) an important way of spreading scientific ideas to the public
- Oersted's disagreement with the accepted idea that electricity and magnetism were unconnected and his attempts to confirm his theory
- Oersted's life as a public lecturer and experimenter.

Context: Work on electromagnets, and the force on a current-carrying wire in a magnetic field.

♦ *Learning outcomes*

Through this topic students can learn:

- about Oersted's experiments on electromagnetism
- about the importance of public lectures in disseminating scientific ideas
- that new ideas can arise when someone disagrees with the accepted theories
- that experiments are necessary to test new ideas
- that agreement on a new idea leads to many other discoveries and applications.

♦ *Teaching strategy*

Oersted's work can be readily reconstructed using simple apparatus and so provides an opportunity for students to engage in practical work where they test Oersted's ideas. Some background information is given in Box 8.3.

Box 8.3 *Background information for students about Oersted*

Hans Christian Oersted

Hans Christian Oersted was born in Denmark in 1777. He attended university in Copenhagen where he became interested in science and philosophy. He was attracted to the idea that light, heat, electricity and magnetism were really just different properties of one force. He travelled around Europe meeting scientists and philosophers and learning about new ideas. A group of scientists he was speaking to in Paris demolished the arguments that he had picked up on his travels and made him realise that all ideas must be tested by experiment. He returned to Denmark and earned a living as a public lecturer while also carrying out his own research (Figure 8.1). His lectures became so respected that he was given a position at the university.

Figure 8.1
Oersted performing a public demonstration.

The controversy: In the late eighteenth century Charles Coulomb had measured the force between charged objects and worked out a scientific law. He did the same for magnetic force but decided that electricity and magnetism were unrelated. Most scientists, including André Ampère, agreed with Coulomb. Oersted was convinced that there was a connection. His experiments proved that an electric current flowing through a wire produced heat and, if the wire was very fine, light as well. But he couldn't get his fine wires to show the expected properties of a magnet – that is, until he was preparing for a lecture in 1820.

Oersted wondered whether a compass would reveal the magnetic properties of an electric current. During the lecture he decided to try out his idea. You can do it yourself.

Experimental details

<u>Activity 1</u>
You need a compass, fine wire and a source of electricity (a small battery will do).

What to do: Lay the wire across the compass and connect the ends of the wire to the battery. Watch the compass as you make the connection (Figure 8.2).

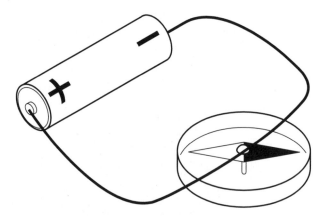

Figure 8.2

Observations: Did you see anything happen?

Conclusions: Oersted saw a small movement of his compass (he was using a very fine wire) but his audience didn't and weren't impressed. He was fairly certain that the electric current in the wire had produced a magnetic force that made the compass move.

He didn't have time to do much research for a few months but then he set to with his compass to plot the magnetic field around the wire. You can do this too.

<u>Activity 2</u>

You need a small plotting compass, wire (thicker and stiffer than before), a battery or power pack (no more than 6 V), a square of stiff card and a clamp stand.

What to do: Make a hole in the centre of the card and push the wire through. Clamp the wire vertically and hold the card horizontally. Connect the ends of the wire to the battery or power pack and switch on. Note the direction of the current in the wire (from + to −). Place the compass on the card and move it around the wire marking the direction of the needle (Figure 8.3).

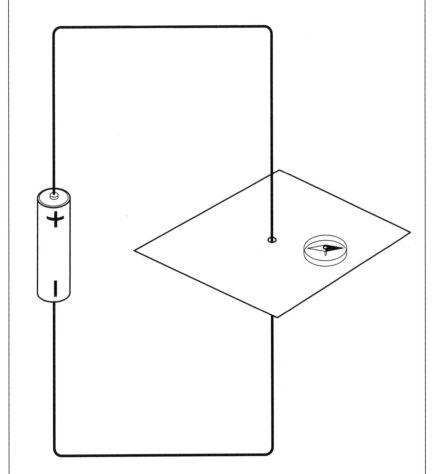

Figure 8.3

Observations: What pattern do your markings make on the card?

Conclusions: What shape is the magnetic field around a single straight wire carrying a current? Oersted found that the magnetic field was in a circle around the wire and devised the 'screw rule'. Find out what this rule is.

Consequences

News of Oersted's discovery created a lot of interest. Almost immediately, François Arago invented the electromagnet and Ampère, quickly changing his ideas, measured the magnetic force between two wires carrying currents. Michael Faraday was inspired to study electromagnetism and soon invented the electric motor and later discovered how magnetism could produce an electric current. These ideas were the foundation of the electricity industry, which now generates electricity in huge power stations for many different uses.

Points to discuss

The career of a public lecturer – discuss the need to be a clear speaker and competent demonstrator familiar with new ideas. Audiences were keen to learn about new ideas and wanted spectacular displays of new discoveries. Where have students seen scientific lecturers at work? They may have been to a Faraday lecture named after Michael Faraday, who was a prodigious public lecturer, or they may have seen the Royal Institution Christmas lectures on television.

Oersted's ideas on electricity and magnetism – these were very much the minority view until 1820 (they were derived from the philosophy of Kant). Ampère, later in the forefront of electromagnetism, was one respected scientist who didn't think that there was a connection.

Oersted had learned his lesson about doing experiments during his European travels but it doesn't seem to have stopped him trying out the compass experiment for the first time in public; he was so convinced that he was right. Was this risky or not?

After doing the experiments, discuss the consequences of the announcement of Oersted's discovery: the rapid sequence of discoveries from Arago, Ampère, Faraday and others and the eventual effects on society through the everyday availability of electricity on a massive scale.

♦ *Other sources*

You can find out more about Hans Christian Oersted and other electromagnetic pioneers in:

Daintith, J. & Gjertsen, D. (ed.) (1999) *A Dictionary of Scientists*. OUP, Oxford.

Ellis, P. (ed.) *breakthrough*. Photocopiable pupils' materials published by PRE*text* (see Other resources, page 155).

Millar, D., Millar, I., Millar, J. & Millar, M. (2002) *The Cambridge Dictionary of Scientists*. CUP, Cambridge.

 www.teachingtips.co.uk (secondary/science/scientists of the month archive)

8.3 Mendeleyev and the Periodic Table

This topic is suitable for use with older secondary students. It allows them to look at:

- how scientific ideas are presented, evaluated and disseminated
- how scientific controversies can arise from different ways of interpreting empirical evidence
- the power and limitations of science in addressing questions, including uncertainties in scientific knowledge.

Context: Work on the Periodic Table.

♦ *Learning outcomes*

Through this topic, students can learn about:

- the Karlsruhe Conference of 1860 and how Cannizzarro's subsequent work provided Mendeleyev with the vital information he needed
- the reluctance of chemists to accept any order of the elements; the importance of supporters of Mendeleyev in overcoming objections
- the power of Mendeleyev's theory in predicting new elements, accepting unexpected elements (Group 0), and its revision in the light of discoveries of atomic structure.

◆ *Teaching strategies*

Two activities are suggested: text-based work researching the historical development of the Periodic Table, and a game to reconstruct Mendeleyev's work ('Being Mendeleyev'). Students have the opportunity to empathise with Mendeleyev by applying his principles to some 'unusual' elements.

Research activity

Students should research the background to the story (see Other sources, page 152) – in particular the Karlsruhe Conference, Mendeleyev himself, and the discovery of new elements that fitted into Mendeleyev's table. Students can be given the questions in Box 8.4; finding the answers will focus their research.

Box 8.4 *Questions about Mendeleyev*

1. What was the main reason for the Karlsruhe Conference in 1860?
2. Why did Mendeleyev need to know the atomic masses of the elements?
3. Why did many chemists dismiss the idea of a pattern to the elements?
4. Mendeleyev himself didn't convince the doubters. What happened to persuade chemists that the Periodic Law was correct?
5. Mendeleyev was not aware of the noble gases when he worked out his table. Why was their discovery in the 1890s not a major problem for Mendeleyev's table?
6. Mendeleyev was not able to answer all the criticisms of his Periodic Table. In what ways is the modern Periodic Table different?

Answers to questions

1. One purpose of the Karlsruhe Conference was to agree on a system of atomic masses. (Agreement was not reached at the conference but shortly after due to Cannizzaro's efforts.)
2. Mendeleyev needed to know atomic masses so he could work out the formulae of compounds from the masses of elements that combined together.
3. There were so many elements being discovered, each with its own unique properties, that it was difficult for many to see the similarities.
4. The discovery of elements that fitted Mendeleyev's predictions convinced most chemists that the Table was useful.

5. The noble gases conveniently fitted into an extra column in the table and so did not disturb Mendeleyev's arrangement. (There was the small problem of argon having a slightly higher atomic mass than potassium.)
6. The elements are in order of atomic (proton) number. There are now 18 groups plus the lanthanide and actinide series so that each element occupies a unique space in the table.

Game

A game that can be used to reconstruct Mendeleyev's work is given in Box 8.5.

Box 8.5 *Being Mendeleyev*

Mendeleyev knew of over 60 elements which he had to place into some kind of order. To get a feel for what Mendeleyev did, imagine you are in an alternative universe where you are Mendeleyev but with only 20 or so elements known to you; because this is an alternative universe they have different names and symbols to what you are used to. Table 8.1 shows details of these 20 elements. To put them into order you must follow Mendeleyev's Periodic Law.

1. The elements should be arranged in order of their atomic mass to form a table. The lightest element goes at the top left.
2. Elements that have similar properties should be arranged in the same vertical column. (The most important feature is the formula of its compound with 'sourgen', a reactive gas that makes up about one-fifth of the air.)
3. Gaps should be left for elements that haven't been discovered yet.

Mendeleyev was a keen card player, so like him you have the details of the elements and their properties written out on cards. Using the three rules given above, arrange the 20 cards into a table.

For any gaps that you leave, predict what the properties of the missing element are likely to be.

Look at the number of groups you have in your table. Why aren't there eight groups?

Table 8.1 *Details of 20 'elements' that students have to arrange according to Mendeleyev's Periodic Law. (These details should be supplied on cards, one for each element.)*

Atomic mass	Symbol	Name	Formula of souride	Properties
1	Aq	Aquagen	Aq_2S	Flammable gas
7	St	Stonium	St_2S	Soft metal, reacts with water forming an alkali
9	Ma	Mabelium	MaS	Metal, reacts with steam forming an alkaline solid
11	Y	Yawnon	Y_2S_3	Non-metal
12	P	Pepson	PS_2	Non-metal, burns in air, forms many compounds
14	Az	Azogen	Az_2S_5	Non-metal, gas found in air
16	S	Sourgen		Non-metal gas in air, supports combustion
19	L	Lumine	L_2S_7	Non-metal gas, very reactive, forms salts (e.g. NL)
23	N	Natrium	N_2S	Soft metal, reacts with water forming an alkali
24	D	Dolomium	DS	Metal, reacts with steam forming alkaline solid
27	H	Hallium	H_2S_3	Strong metal, reacts with steam
28	Q	Quartzon	QS_2	Semimetal, main element in sand and glass
31	G	Glowon	G_2S_3	Non-metal, burns spontaneously and forms an acid
32	B	Brimstone	BS_3	Yellow non-metal, burns in air to form an acid
35.5	Gr	Greenine	Gr_2S_7	Green gas, very reactive, forms salts (e.g. NGr)
40	Li	Limium	LiS	Metal reacts with water forming alkaline solid
70	Pa	Parisium	Pa_2S_3	Metal, reacts with steam
73	De	Deutchium	DeS_2	Semimetal
78	Lu	Lunon	LuS_3	Non-metal, burns to form acids
80	Br	Brownine	Br_2S_7	Liquid non-metal, very reactive, forms salts (e.g. NBr)

Commentary

Research activity

Mendeleyev's Periodic Law is one of the most important chemical laws, particularly at GCSE level, and yet it is often taken for granted. Until 1860 it was very difficult for anyone to come up with a pattern of the elements. New elements were being discovered frequently, especially after the invention of spectroscopy by Bunsen and Kirchhoff in the 1850s. More importantly, until the 1860s there was no agreed system of atomic masses so the formulae of compounds were frequently disputed. The Karlsruhe Conference illustrates the importance of meetings of scientists for the development of ideas.

Many chemists simply didn't see the need for there to be a relationship between the elements – they were elements after all, which meant that they were distinct and separate substances. Members of the Chemical Society ridiculed Newlands' attempt to draw up a table. Mendeleyev, though, felt that there was a pattern and drew up his list of rules for compiling the table. He was prepared for a battle to justify his ideas and the table itself did not convince everyone. It was the discovery of elements that matched his predictions and the efforts of his supporters that finally brought acceptance.

The discovery of the noble gases, which fitted relatively easily into an extra column in the table, added weight to the law but there were limitations to Mendeleyev's table. There were still a few elements that were out of place (Te and I, Ar and K) and the lumping together of the iron group metals and the strange double groups (e.g. 1A the alkali metals and 1B the coinage metals) could not be explained. Subsequent discoveries of atomic structure produced the modern version.

'Being Mendeleyev'

'Sourgen' is, of course, oxygen – it does not combine with itself so there is no formula, but all the other elements do combine with it and the formulae are provided. Note the pattern of the formulae (X_2S, XS, X_2S_3, XS_2, X_2S_5, XS_3, X_2S_7)

'Aquagen' is hydrogen and according to the formula of its 'souride' it should go in the first group with the next element 'stonium' (lithium) directly below it.

To complete the table with all 20 elements, two gaps must be left. They correspond to:

potassium – a soft metal that reacts with water to form an alkali; the formulae of its 'souride' is M_2S.
arsenic – a non-metal; the formula of its 'souride' is X_2S_5.

Note that there is no group 0 (noble gases) since Mendeleyev did not know of them. Also there are no transition metals; you could point this out to students, to illustrate the problems that Mendeleyev faced.

◆ *Other sources*

Brock, W. (1992) *Fontana History of Chemistry*. Fontana, London (Chapter 9 Principles of Chemistry, p. 311ff).

Daintith, J. & Gjertsen, D. (ed.) (1999) *A Dictionary of Scientists*. OUP, Oxford.

Millar, D., Millar, I., Millar, J. & Millar, M. (2002) *The Cambridge Dictionary of Scientists*, CUP, Cambridge.

Strathern, P. (2001) *Mendeleyev's Dream – the Quest for the Elements*. Hamish Hamilton, London.

Woods, G. (2001) A brief history of the Periodic Table. In Ellis, P. (ed.) *breakthrough*. Photocopiable pupils' materials published by PRE*text* (see Other resources, page 155).

8.4 The DDT story

This topic is suitable for use with younger secondary students. It allows them to look at:

• how science is reported in academic journals, books and the media
• scientific controversies about the effects and effectiveness of DDT
• the influence of context on scientific work, and how this affects the work's acceptance
• the power and limitations of science in addressing industrial and environmental questions.

Context: Work on the impact of humans on the environment.

◆ *Learning outcomes*

Through this topic, students can learn about:

• the development of DDT as a pesticide and the beneficial results of its use during and after the second world war.
• the work of Rachel Carson in highlighting the detrimental effects of DDT on the environment
• the arguments for and against the use of DDT today

- the role of the media, government, commerce and environmental groups in the DDT story.

◆ *Teaching strategies*

This topic lends itself to research by pupils using the internet. They can also present the DDT story as a drama or role-play exercise.

Internet research

Box 8.6 *The DDT story*

DDT was one of the first synthetic pesticides to be manufactured and used on a large scale. At first it was hailed as a wonderful benefit to mankind, but now it is banned in many parts of the world. In this unit you will learn about the people involved in the DDT story, why it was widely used and why it fell out of favour. The DDT story may be history but you will see that it still has an important part to play in today's discussions about the uses of pesticides.

Use the websites listed to research the DDT story. In particular, find the answers to the following questions:

- Who discovered that DDT was a useful pesticide?
- What was DDT used for and why did its use increase?
- Who discovered the problems with DDT and how were the public notified?
- What was the result of the announcement of the dangers of DDT?
- Is DDT still in use today? If so, where?
- What reasons are there for and against continued use of DDT?

Websites

http://www.chem.ox.ac.uk/mom/ddt/ddt.html
molecular structure and history of DDT

http://www.nobel.se/medicine/laureates/1948/muller-bio.html
biography of Nobel Prize winner Paul Muller

http://www.chem.duke.edu/~jds/cruise_chem/pest/pest1.html
the reasons for using DDT

http://www.rachelcarson.org
site dedicated to Rachel Carson

http://www.uneco.org/silent_spring.html
excerpts from Rachel Carson's book *Silent Spring*

http://members.aol.com/rccouncil/ourpage/index.htm
organisation dedicated to continuing Rachel Carson's work;
includes a biography

http://whyfiles.org/coolimages/captions/eagle.txt
news of results of ban on DDT

http://www.seattletimes.com/news/nation-
world/html98/ddtt_19990829.html
comment on another effect of the ban

http://www.findarticles.com/cf_0/m1200/1_158/
63692741/p1/article.jhtml
a summary of the issues

 Students can research the internet to build up their knowledge and understanding of the DDT story given in Box 8.6 and to develop their awareness of the issues.

Drama or role play

Students can be asked to devise a short play or debate looking at the DDT story. They may want to develop the following roles:

Paul Muller
A farmer
A naturalist (bird watcher)
Rachel Carson
Pesticide manufacturer(s)
Government official(s)
Environmental activist(s)
Parents of child with malaria
(plus a cast of mosquitoes, fish, birds, etc.)

Commentary

The DDT story is one of success for science in solving one problem but failure to predict all the outcomes of a worldwide experiment. The story spans 70 years of the twentieth century and carries many lessons for us today, particularly the reaction of industry and governments to unexpected and unwanted effects. A similar story has occurred in the use of leaded petrol, CFCs, phosphate detergents and many other triumphs of modern technology. As well as exploring the present-day arguments for and against the use of DDT (it is still a lively debate), looking at the history of the story can give students an insight into how science in the hands of ordinary people can result in benefits and costs to themselves and the environment.

◆ *Other sources*

Carson, R. (2000) *Silent Spring*. Penguin, Harmondsworth.

Other resources

breakthrough is a photocopiable resource, published in three packs. Each pack looks at aspects of the history of science that are relevant to the secondary science curriculum, and suggests suitable teaching approaches. Some articles look at the work of practising scientists, and others suggest places to visit. *breakthrough* is available from PRE*text*, Brecon Cottage, 33 Newbury St, Wantage, OX12 8DJ (pretextpub@aol.com). PRE*text* also publish a list of books on the history and nature of science, which they can supply at reduced prices.

9 Ideas and evidence: Contemporary questions

Tony Sherborne

At school, science can appear to be about topics that were thrashed out many years ago. Flick through many textbooks and you will see few contemporary faces – the 'dead white male' syndrome. To be fair to textbooks much of the science covered by the curriculum is well-established, and was worked out long ago. However, we live in a world that is changing rapidly, largely as a result of technological change based on recent scientific discoveries.

Our students can be excited by this, and we hope that they will be, for several reasons:

- Some of our students are the scientists of the future, and we want them to engage with the latest ideas.
- Modern science and technology will be crucially important in solving some of the biggest problems facing the world today.
- As citizens, students will have an influence over what controversial science is carried out – and they need to learn how to interrogate the media if they are to gain an accurate picture of the issues.

These aims lie behind some of the requirements of the ideas and evidence strand of the curriculum. This chapter suggests some approaches to tackling them.

9.1 How scientists work today

This topic is suitable for use with younger secondary students. It allows them to look at:

- the ways in which scientists work today, including the roles of experimentation, evidence and creative thought in the development of scientific ideas.

Context: Work on cells and their functions. (I have assumed that students have already met ideas about how cells divide.)

♦ *Learning outcomes*

Through this topic, students can learn that:

- an experiment is a way to answer a scientific question, or test a new idea
- a new idea is the result of a scientist's creative thought
- a new idea must be backed with evidence in order to become accepted.

It can also help them to develop the skills to:

- use skimming, scanning, highlighting and note-taking
- select and synthesise information to meet needs.

♦ *Background*

In primary schools students are introduced to the importance of experiments, evidence and creative explanations. Students concentrate on their own investigations, but also study famous scientists of the past. Younger secondary students should also explore the work of modern scientists.

Contemporary case studies may present additional challenges to students. In particular, the concepts behind more recent scientific ideas are often too demanding for students of this age. The challenge for the teacher is therefore to simplify the ideas enough for these students, as well as making it relevant to the curriculum. However, to avoid contemporary science examples would be to do students' interests a huge disservice. It is stories such as these, from the frontiers of science, that offer students the inspiration that could help them to see their future in the subject.

Case study 1 in this section focuses on a scientist whose work has recently been honoured with the greatest accolade, a Nobel Prize. Sir Paul Nurse, who heads the Imperial Cancer Research Fund in London, won the prize for medicine in 2001. He and his colleagues discovered the gene that seems to control cell division in all living things. Since faulty cell division in humans is what underlies cancer, the discovery could lead to new treatments. Paul's work also illustrates all the different elements of the ideas and evidence statement about how scientists work:

- **Creative thought:** Paul discovered the control gene in yeast, and wondered whether it could also control cell division in other living things.

- **Experiments:** Paul found the gene that controlled division by treating cells with chemicals to alter the DNA, and observing whether the cells continued to divide as normal.
- **Evidence:** Paul's results showed that, once a particular gene was blocked, a cell started to divide out of control (like a cancer cell). And this gene also worked as a 'control gene', when substituted into other animals.

◆ *Teaching strategy*

You could use a 'research and write about' approach for students to explore Paul Nurse's work, and deepen their understanding of the elements in the scientific process.

Case study 1: Cancer genetics

Information on the task is given in Box 9.1 (note that the Nobel Prize criteria have been adapted and simplified for this task).

Box 9.1 The work of Paul Nurse

Figure 9.1
Paul Nurse, a UK scientist has won the Nobel Prize for medicine

A discovery must meet four criteria to win this Prize:

1. It should help scientists to develop new treatments for disease.
2. It should involve creative thought.
3. It should involve experiments.
4. It should be backed up with evidence.

You will research and write a one-page article about it for a news website, aimed at young people.

Your article should contain four paragraphs. Each paragraph should explain how Paul Nurse's work meets one of the criteria. For your research, you will use websites, information sheets, videos or experiments.

This assignment is a combination of two teaching approaches. The research is an *active reading* task; producing the article is an *active writing* task. Active reading tasks encourage students to search actively for, categorise and evaluate information from text. The idea is that, by interrogating the text, they will understand the material better. Active writing tasks add the elements of purpose, audience and style, to students' writing in science. Writing in general will help students to clarify their ideas, in this case about evidence, experiment and creativity, in the process of articulating them on paper. If writing is given a real and different purpose from 'writing for teacher' this adds motivation and variety. Active writing is also highly effective in developing students' literacy, communication and thinking skills. This approach will give them practice in:

- coping with new scientific terminology while researching (literacy)
- categorising information found (thinking skills)
- selecting and summarising relevant material (communication skills)
- writing coherent sentences and paragraphs (literacy).

Several techniques are appropriate to support students in developing these skills:

- providing a glossary of science terms that students might meet in their research
- a clear structure for the article, with sections
- a writing framework, with sentence beginnings.

 Both the research and writing tasks could also be done using ICT. Pupils could research using particular websites, and could prepare their article using a word processor.

Age-appropriate published materials about Paul Nurse's work form part of *ACCLAIM*. This is a set of resources from the Royal Society and the Centre for Science Education, Sheffield Hallam University. Some of the *ACCLAIM* resources are freely available via the website. Others may need to be purchased, like some of the activity sheets and the video (although it was originally broadcast free to air on 4Learning).

1. The active research task

The activity will be easier if students make a plan for their research notes. With a page for each section, they can then write relevant notes (or cut and paste them) directly into the correct section when they find them. Here are some sources for the information needed in each section:

Criterion 1: new treatments

ACCLAIM activity 1: the story behind the press release. An active reading task to give pupils an overview of Paul's research.

ACCLAIM video: This includes the connection between Paul's work and cancer.

Criterion 2: creativity

ACCLAIM website: The text of an interview with Paul Nurse, where he answers questions on creativity.

Criteria 3 and 4: experiments and evidence

ACCLAIM activity 4 (available on website): Looking at cell division. A simple look at the experiments Paul carried out on yeast, and the conclusions he drew.

ACCLAIM activity 5/6 (only in pack): Students can simulate Paul's work, by growing some of the yeast cells for themselves.

ACCLAIM website: Links to articles describing Paul's work in more detail.

2. The writing task

This task is really made up of three parts: selecting which information to include, ordering it into paragraphs and then writing the sentences. A writing frame can provide support for students in these processes. Some sentence beginnings that will give pupils experience of writing in different ways are given in Box 9.2.

Box 9.2 *Writing frame*

Paragraph 1: new treatments

(a 'reporting frame' – a general description, which gradually becomes more technical)
The science topic that Paul was researching is . . .
The disease related to his discovery is . . .
Paul's research could lead to . . .

Paragraph 2 – creativity

(a 'persuasion frame' – presenting the case, and then giving arguments for it)
The new idea Paul thought up is that . . .
The reason this is a creative thought is . . .
Paul thinks up new ideas because . . .

Paragraph 3: experiments

(a 'procedural frame' – an ordered description of the method)
Paul did experiments on . . .
The first step of Paul's experiment was . . .
Then he . . .

Paragraph 4: evidence

(an 'explaining frame' – sets out a logical argument)
The conclusion of Paul's experiment was . . .
This is evidence for Paul's idea that . . .

◆ *Alternative topics*

You could use a similar approach to study many other leading scientists – arguing for why they should win the prize in future. This could become a 'balloon debate'.

◆ *Other sources*

The pack *ACCLAIM: Exploring the Lives of Leading Scientists*, available from the Centre for Science Education, Sheffield Hallam University.

 ACCLAIM website: www.acclaimscientists.org.uk

You will find accounts of the work of other contemporary scientists, written by the scientists themselves, in *Catalyst*, the GCSE Science Review, published by Philip Allan Updates.

9.2 Mobile phone safety

This topic is suitable for use with older secondary students. It allows them to look at:

• how scientific controversies can arise from different ways of interpreting empirical evidence and models based on this evidence.

Context: Work on the electromagnetic spectrum, its use and dangers.

◆ *Learning outcomes*

Through this topic, students can learn to:

• recognise situations where there is scientific disagreement over a finding
• identify, or suggest, at least two different explanations for a finding.

◆ *Background*

By this stage of their secondary science course, students will have looked at the difference between 'evidence' and 'interpretation'. They learn that interpretation is something that goes beyond the results themselves, and can require creative thought. Now students can understand how findings can often be interpreted in several different ways; and that this leads to scientists taking different views – a scientific controversy.

Teaching about controversies in science can be challenging, both for students and for teachers. Students are used to finding a 'right answer' – and here there isn't one. Teachers are also likely to feel on less secure ground, with topics where there are often few facts, and where the material may be unfamiliar.

For the case study in this section, I have chosen a context that students will find relevant, mobile phone safety. There have been suggestions that the microwave radiation that mobiles emit can cause certain brain cancers. The research, which compares the incidence of brain tumours among mobile phone users with that for non-users, is probably inconclusive. (The position may have changed by the time you read this!) Some scientists have interpreted the findings as showing an increased risk. Others say that there is no significant risk. In other words, the evidence can be interpreted in more than one way.

◆ *Teaching strategy*

The teaching approach you could use is 'small group discussion' (SGD). It is ideal for exploring the idea of controversy because, in a discussion, students can take different points of view and experience the disagreement at first hand. Such experience may make it easier for them to accept that there is not always an accepted answer.

SGD has several other important benefits:

- it allows students to express their opinions in science, maximising each individual's contribution
- it develops key skills such as communication and working with others
- it enables students to clarify their ideas by talking
- it allows students to be creative, and learn from each other
- it is an effective platform from which to launch a whole class discussion.

However, SGD is a strategy that many science teachers do not use often. It certainly has challenges, such as:

- keeping students focused on the task of discussing the science
- getting students to talk
- making sure that some learning has taken place.

To help overcome these, Case study 2 uses the following techniques:

- it provides an 'agenda' for the group to work through – to give purpose and structure to the discussion
- it prescribes a clear outcome – what the group has to produce by the end – and creates an expectation of 'reporting back' to the class
- it splits students into 'expert groups' before the discussion – to prepare what they are going to say.

Case study 2: Mobile phone safety

ICT This activity is taken from the pack *Ideas and Evidence*, developed by the PRI. The unit on mobile phone safety can be downloaded from the website (see page 166).

A worthwhile discussion needs building up to, so the activity is structured into several parts:

Engagement
This first aim is to get students interested in the discussion question and refresh their memories on the relevant science. According to a recently published survey from PRI, 10% of teenagers use their phones for at least 45 minutes a day. This level of usage is against current government and scientific advice. If there turn out to be health risks, they are likely to be more serious for young people, whose developing brains are more vulnerable.

To get a better sense of the potential dangers, students can read a newspaper report. The unit provides an adapted article called 'Microwaves gave my husband a tumour'.

Preparation
The question to be discussed is whether using a mobile phone can harm your health. The discussion is set within the context of a 'TV debate', where various experts argue their points of view. The activity gets students to take on the viewpoint of one of the experts (e.g. a spokesperson from the mobile phone industry). Students are given a card with several pieces of evidence that the expert might use to argue his/her case (e.g. the sample card in Figure 9.2). The students will take turns to present these points in the discussion, as a springboard to further debate. In the process they will see that different experts are often using the same pieces of evidence to support opposing points of view.

Professor D Gupta: Leader of Research Project looking into mobile phone safety

Scientists in Sweden looked at 200 people with brain tumours. They found no evidence that using a mobile phone increased the chance of having a tumour. But, mobile phone users were over twice as likely to have a tumour near their phone ear than in other parts of the brain.

Scientists in the US did similar research, using 450 people with brain tumours. Again they found no evidence that using a mobile phone increases the chance of having a tumour. But there were more than twice as many mobile phone users who had a particular type of tumour (called a neurocytoma) compared with people who did not use mobile phones.

We must be careful. Firstly, the number of mobile phones is increasing at an incredible rate (see diagram). So even a tiny increase in the risk of brain cancer could cause many more deaths across the world.

Secondly, brain tumours can take many years to develop. Mobile phones have not been around for very long. So it might be a long time before we find out whether mobile phones cause brain cancer or not.

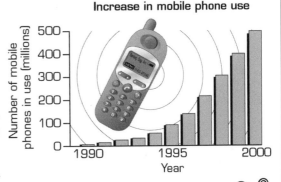

Increase in mobile phone use

Figure 9.2
A sample card from the Ideas and Evidence *PRI pack unit.*

In order to talk confidently, students need to prepare. They should become familiar with the evidence and understand the arguments on their card. A good way to do this is to split them into groups of the same character (e.g. a group of Professor Guptas). This technique is known as 'expert groups'.

Students are then given 5–10 minute to prepare together in their expert group. They can be directed to:

- highlight the key evidence for the point of view or make notes, or both
- highlight words or phrases they don't understand, and help each other
- practise summarising the main points to each other
- devise questions they could be asked.

Discussion

The ground is now paved for an effective discussion. Groups are given an agenda that structures the debate into three parts:

1. Experts present their arguments and evidence.
2. People can ask each other questions.
3. A spokesperson notes down a list of 'points they agree on' and 'points they disagree on'.

Reflection

It is vital to allow time afterwards for students to reflect on what they have learnt. It is all too easy for them to come away with only half the story – just that there is a disagreement. A good approach here is to have a plenary discussion. This could start with groups reporting back their lists. From here the teacher can focus on one of the research findings, and elicit the idea that the evidence can be interpreted to support either point of view.

The aim is for students to understand that the process of science does not always generate clear-cut answers. Just like other people, scientists often hold particular points of view. They may use the evidence in one of several possible ways, or even use the evidence selectively, in order to support their view.

◆ *Alternative topics*

To broaden students' understanding of the ideas and evidence objective, here are some other scientific controversies which could be tackled via small group discussion:

◆ 'What killed the dinosaurs?' – different explanations include an asteroid and volcanic eruptions.
◆ 'Are vaccines like MMR safe?' – examining the controversial claim that it causes autism.
◆ 'Is there life on Mars?' – does the famous meteorite really contain fossilised bacteria?

◆ *Other sources*

Ideas and Evidence resource pack (2001), by the Pupil Researcher Initiative. Collins Education, London.
The unit 'How safe are mobile phones?' is available from: www.shu.ac.uk/pri

9.3 How the media influence science

This topic is suitable for use with older secondary students. It allows them to look at:

* how scientific ideas are disseminated.

Context: Work on how animals are adapted for survival in different environments. (The general approach of this case study could be adapted to almost any topic.)

◆ *Learning outcomes*

Through this topic, students can learn that:

* the media influence the science they report by choosing stories selectively, and can affect the points of view that people take on the issues they report
* there are several common types of bias in a science news story, such as sensationalism, overcertainty and trivialising the science.

◆ *Background*

Students will have looked at aspects of the media in the lower secondary school, as part of learning about the role of evidence. They practise looking for evidence to support arguments made in the media, and in distinguishing facts from opinions.

Older secondary students should develop a broader understanding of the role of the media. The learning outcomes listed above reflect the expectations of the GCSE exam boards.

These learning outcomes are surely an important part of 'science literacy'. We want students to become citizens with the skills and confidence to engage with science-based issues, and to reach informed opinions. If the media are the main source of science information, and they tend to distort science, then it is crucial that students learn how to interpret critically the science they see or hear about.

Students may find that learning to interpret the media is more difficult with science than with other subjects. One reason is that it is not easy to assess how accurately ideas are being reported when the ideas themselves are often hard to understand. Another is that scientific findings tend to sound believable and convincing. It is only by examining the methodology behind the research that you can tell whether the finding or the media's 'soundbite' is accurate – and this detail could well be lacking (see also section 7.3 on pages 126–133).

♦ *Teaching strategy*

One approach to learning to interrogate the media would be to study newspaper articles, comparing the media's account with the original. Case study 3 in this section looks at a different approach which is based on students actually experiencing the process of 'disseminating science' for themselves.

This type of teaching strategy is called a simulation – students model the real situation of being in a TV newsroom, presenting science. Simulations give students concrete experiences, and can be highly effective in learning 'soft' concepts such as 'the influence of the media'. They have many other benefits:

- they can make science seem more relevant, by modelling a real situation
- they can generate high levels of motivation
- they can increase students' confidence
- they provide differentiation: being open-ended, they enable students to contribute their strengths
- they develop key skills: communication, problem solving and working with others
- they develop thinking skills: handling information, reasoning, creativity and evaluation.

With all these benefits, it is not surprising that simulations can be more time-consuming than traditional methods, both to act out (students need time to 'think themselves into the situation'), and to prepare.

Simulations also pose new management challenges to the teacher, because during the action the students are in control (and hopefully not 'out of'). Also, since simulations are realistic tasks, they are inevitably demanding for many students.

To minimise potential problems and maximise the benefits, attention needs to be paid to each stage of the simulation:

- **Planning:** What information will students need? What jobs will they do?
- **Briefing:** What are the structure, rules, timing and outcome of the activity?
- **Action:** Will you intervene at all? How will you monitor progress?
- **Follow up:** How can you get students to reflect on what they have learnt?

Case study 3: Getting into the media

This case study draws on a published simulation. It is called 'Getting into the media', and is part of the *Ideas and Evidence* resource pack, developed by the PRI.

In the simulation, students take on the job of a news journalist working for a 'popular' TV channel. They have two tasks:

A To choose the most 'newsworthy' story from a choice of three. This task teaches students how the media influence science by selectively reporting it.

B To produce a 60-second TV news item, from an information-filled press release. The idea is to create the same constraints of time, and to keep viewers watching, as in real TV. This forces students to twist the science they are reporting – teaching them how and why the media can also be biased.

Almost any science topic could be chosen for this stimulation. This case study focuses on 'Advanced materials', which links to various parts of the curriculum. Here is how each stage in the simulation works:

Preparation
Students will need several materials to help them in their tasks:

- **Task A:** Guidance for judging which is the best story – a checklist of features of a good TV news story. These features, called 'hooks', are: 'emotions', 'something surprising', 'something life-changing', 'important new fact' and 'great pictures'.

- **Task B:** Press releases – the raw material, with the story and background science. There are three press releases in the published simulation, each about different advanced materials. One is shown in Figure 9.3.

Media release

'Penguins help scientists design super-clothes'

Imagine a jacket that you could scrunch up and fit into your pocket, but which keeps you warm on the iciest of days when you're wearing it. This new material could transform our winter wardrobe. Instead of big chunky coats, we could have slim-line jackets – keeping us warmer than ever before.

Dr Jules Pincent at Redstone University has studied how penguins stay warm in the Antarctic to produce a sample of very thin but highly insulating material, which could make such a jacket.

He said: '*nature is often a good starting point for trying to improve our own capabilities – whether in clothing, flying or whatever.*'

Penguins manage to stay warm even in temperatures of –30°C. Dr Pincent has found that they have small curved feathers which lock into place in cold weather, trapping air pockets underneath. Air, a poor conductor of heat energy, reduces the penguins' heat loss. The thing that improves the insulation of the feathers even more is what is underneath them. These are bundles of tiny fibres – only 5 millionths of a metre thick – which are scrambled up like a ball of tangled wool, and trap hundreds of tiny pockets of air underneath a single feather. So the air is completely trapped in place. It's very hard for heat energy to escape.

In warm weather, the penguins can pack down their feathers, crushing the tiny fibres flat. This loses the 'blanket effect' and means the penguins don't overheat.

Dr Pincent uses a similar approach for his new clothing material. It contains layers of curved material, where each curve has tiny fibres underneath. It has very good insulating properties. The tricky bit is finding fibres that are thin enough. He is still looking for something as thin as the penguins' fibres.

Comment from clothing designer Bjorn Blodvick: '*We already have fantastic materials to keep us warm in cold weather, like fleeces and breathable jackets. What's new about Dr Pincent's material?*'

Media contacts
For more information about this story, visit: **www. shu.ac.uk/pri**

HOME

© Pupil Researcher Initiative & Collins Education

Figure 9.3
Penguin clothes press release (from the unit 'Getting into the media' from Ideas and Evidence *resource from PRI).*

- **Planning sheet:** the parts of a news item (newspaper introduction, reporter's summary, plus interviews), what should go in each part, and ideas on creating impact.

Students can work in groups of four or five. Four will be involved in presenting the news item: the newsreader, the reporter and up to two people interviewed.

Briefing
Here you explain the structure and rules to the students.

Task A: Outcome: the most newsworthy story – the TV news has only one slot for a 'science story'.
Rules: read each press release. Look for the 'hooks' it has. Choose the most newsworthy story – the one with the most hooks.

Task B: Outcome: a 60-second presentation. This airtime constraint is realistic and forces pupils to be very selective in which information they present.
Rule: only the 'best' story from the class 'journalists' will be broadcast.
The competitive angle forces students to focus on producing a 'sexy' news item, and in the process sensationalise the story or trivialise the science.
Rule: the items will be presented and judged. Ratings are everything for this channel. The item chosen will have the most 'hooks' (as in task A).

Action
The exciting part of the action is when students transform the press release into a news item to broadcast. They will be engaged in writing scripts, devising characters to be interviewed and in rehearsing the piece. Our experience from trials of this topic is that the 'winning item' is likely to distort the original story in some of the following ways:

- overselling the benefits of the material and ignoring its drawbacks
- suggesting with certainty that the material will work
- oversimplifying the science, or presenting it inaccurately
- hyping up the story's importance, by using controversy between scientists.

Follow-up
A discussion would be a good way to get students to reflect on what they have learnt. For task A, students need to remember the criteria they used in selecting their story. They can then make the leap to the real-world media, where the stories you see on TV are not necessarily the most important ones – and sometimes even not good science.

For task B, the discussion would aim to make explicit as many ways as possible (like those above) where the students, acting as the media, have been 'biased' in presenting science.

◆ *Alternative topics*

 To find press releases on other science topics, which you can readily adapt, visit the website: www.eurekalert.org/pubnews.php

◆ *Other sources*

Ideas and Evidence resource pack (2001), by the Pupil Research Initiative. Collins Education, London (see Chapter 6, page 114, for more detail).

10 Teaching citizenship through science

Peter Campbell

The National Curriculum was extended in September 2002 to make citizenship education compulsory, though flexible in its implementation. State schools in England must now demonstrate that their students are taught, for example, about the machineries both of government and of justice, as well as connections between local activities and global challenges. Students must also be taught skills associated with enquiry, discussion and participation, appreciating the need for mutual respect among people of different backgrounds and views.

In a single century, extraordinary successes of science and technology have enabled people to change the world. This changed world has come to rely upon a significantly increased number of the population working in science-related employment, some 60 million. Indeed, science-related issues play such an important part in contemporary life that citizenship education must surely include them.

10.1 What citizenship education is about

During their years at secondary school, young people's horizons broaden and so does their curiosity about the world. During these years it is likely they will encounter, either among themselves or in the mass media, many viewpoints that conflict with those they have taken for granted. Starting from students' interests, perceptions and viewpoints, the responsive teacher will have many opportunities to develop their ability to consider and debate live issues.

What citizenship concepts should students be developing? They need to understand that rights and responsibilities are usually flip sides of the same coin. Your right to life is my responsibility for your safety and security. At the extreme, murder is a violation of another person's right to life. Students can learn to empathise with others, and understand how conflicting viewpoints arise. A farmer's livelihood can be threatened by animal disease, and he accepts the enforcement of hygiene standards on farms because these provide protection for all. But if the supermarket consumer wants cheap food doesn't that create pressure to cut corners? A basic requirement for public debate in a liberal society is that tolerance and mutual respect should exist alongside the expression of divergent views. Negotiating a way forward can prove much harder; compromise often satisfies none of the parties to a dispute. 'Balancing' views in an argument might require that unheard voices are identified too and expressed.

Students should develop their ability to weigh up viewpoints. They might weigh up the scientific evidence associated with an issue, for example what risks are associated with MMR vaccine? To answer that question, they will need to consider what we mean by 'risk', how intuitively and frequently we calculate risks and how human psychology can lead us to exaggerate or underestimate risks. Students discover that scientists do not always agree among themselves, and yet choices must be made. Will my toddler be vaccinated? Isn't she better protected when most people's children have been vaccinated?

Students might also weigh up values and beliefs underlying judgements, appreciating that we argue within ourselves sometimes. How does my preference for eating exotic fruit brought into the country by air freight square with my concern for the environment? Too easily, environmental concern expressed in the classroom is simply put aside when concerns for friendship, fashion and belonging demand something else. Let's face it, the social messages in a society greedy for energy are hard, if not impossible, to resist. Something as simple as how students get to school each day can point to the need for thinking about sustainable development.

In a world that is preoccupied with commodities and that sometimes seems to encourage selfishness the very notion of common interests may need to be justified, even those as basic as clean water and clean air. This is true at every scale, from globally to within the school itself. Students are learning how the world works – not just its machinery but also the unavoidable tensions that the machinery tries to manage.

10.2 Controversy and discussion are central

By its nature, citizenship education is only in small part a matter of instruction. Generally, teachers need to avoid the role of scientific 'expert' and instead maintain procedural authority, structuring student discussion and research. Strongly held views are not necessarily problematic, but if you are not careful then extreme views might be expressed that are offensive to others in the class. Principles underlying discussions of all types need to be clearly established. A successful citizenship lesson is one that leads not to classroom consensus but to thorough exploration of an issue, protecting diverse student views.

Many science teachers will find this unfamiliar role challenging. With a standard class of 28 girls and boys, taught in a mixed-ability group, there are many possible pitfalls. Teaching materials cannot be purely text-based or poor readers will be disenfranchised and perhaps become hostile. Full class discussion might allow some pupils to dominate, leaving others passive if not disaffected. The English or humanities teacher has numerous strategies for overcoming difficulties such as these: activities that encourage close reading (sequencing, text marking and text restructuring) and structured discussion strategies (pair talk, pairs to fours, envoys, jigsaw and consequence mapping). Newspaper cuttings can be deconstructed using these methods, without forgetting that the lower reading age for tabloids makes these more accessible to many students than are quality newspapers. If the new challenges of citizenship education mean that time is made available to bring teachers from different subjects together more often, to share pedagogic approaches and knowledge, it will be a very good thing.

Other helpful strategies include the use of images as stimulus material: video clips, photographs and cartoons. Images carry powerful messages. A photo of pet calf Phoenix in the news, for example, changed many minds on the matter of culling animals to control foot and mouth disease. Students generally enjoy reading aloud, and many enjoy acting. Drama, with its intrinsic human interest, can also be used to motivate discussion. In the right circumstances, students enjoy expressing their own views, both orally and in writing. For this reason, citizenship education can make a major contribution to the development of language and literacy. By using both the citizenship and science words in contexts meaningful to students, their understanding of new concepts is encouraged.

10.3 What citizenship education can include but is not about

Although there is no single correct answer to a social problem, there can be wrong answers. A proper concern for ideas and evidence arises naturally from considering science in its social context. Whether it is a confrontation in court over forensic evidence, or a social attitude to the spread of AIDS that fails to acknowledge the ecological strategy of the virus, false outcomes occur when information is lacking or evidence is unreliable. There is an important message here about the nature of science: students too often see science as a body of fact and too seldom appreciate that it can be characterised as a battle between conflicting ideas, won sometimes by subtle evidence.

So also the history of science can be used to shed light on contemporary issues. Dr Christiaan Barnard carried out the first heart transplant in South Africa in 1967, bringing him fame and glory. In his autobiography, he used the following much-quoted metaphor to explain the choice faced by his patients (*Christiaan Barnard: One Life*. Macmillan, New York):

> For a dying person, a transplant is not a difficult decision. If a lion chases you to a river filled with crocodiles, you will leap into the water convinced you have a chance to swim to the other side. But you would never accept such odds if there were no lion.

Without denying his undoubted surgical skills, transplanting a heart became possible only because medical science at that time understood the human immune system well enough to solve the problem of organ rejection. Today, because of cultural and religious beliefs, many people still do not accept that human organs should be transplanted and an ongoing campaign of public education is needed to recruit donors.

Once everything taught at school has fallen away, the public understanding of science relies heavily on the role of media. For students this makes an interesting study in itself. They can examine media coverage of contemporary science-related events, checking to what extent the public is provided with useful background science and to what extent it simply feeds public hysteria, as the American media famously did with apples and the pesticide Alar during 1989. If students learn to be critical of media science coverage the habit might usefully serve them for a lifetime.

10.4 Summary

Research evidence shows that what happens in science classrooms during Key Stage 3 is crucial to post-16 subject choices. The teaching of citizenship through science offers a major opportunity to liven up some science lessons and engage students.

Other resources

The primary source of stimulating resources continues to be practising teachers, who have tried out new ideas in their own classrooms. This should inspire other teachers with confidence that they can do the same. There are many excellent published resources to support the teaching of citizenship through science:

- The ASE SATIS project (Science and Technology in Society) has published several series of booklets; contact ASE Booksales, College Lane, Hatfield, Herts AL10 9AA.
- A derivative of these is *The World of Science*, a student's book and photocopiable resource pack (John Murray, London).
- *CHARIS Science 11–14*, from The Stapleford Centre, The Old Lace Mill, Frederick Road, Stapleford, Nottingham NG9 8FN.
- Resources specifically aimed at teaching citizenship in a science context have been developed by the ASE in collaboration with the Wellcome Trust. These are published on the Science Year CD-ROM: *Can we? Should we?*, which is available from ASE Booksales (details above) or via http: //www.sycd.co.uk

- Units 20 *What's in the Public Interest?* and 21 *People and the Environment* of the QCA Scheme of Work for Citizenship at KS3 can be found at http://www.standards.dfes.gov.uk/schemes

Index

INDEX